THE PERMAFROST ENVIRONMENT

Stuart A. Harris

CROOM HELM
London & Sydney

Croom Helm Ltd, Provident House, Burrell Row,
Beckenham, Kent BR3 1AT

Croom Helm Australia Pty Ltd, Suite 4, 6th Floor,
64-76 Kippax Street, Surry Hills, NSW 2010, Australia

British Library Cataloguing in Publication Data

Harris, Stuart A.
The permafrost environment. – (The Croom Helm natural environment; Problems and management series)
1. Civil engineering – Cold water conditions – Frozen ground
I. Title
624'.0911 TA713

ISBN 0-7099-3713-X

Printed and bound in Great Britain by
Biddles Ltd, Guildford and King's Lynn

CONTENTS

PREFACE

While teaching a course on the use of permafrost during the last five years, it quickly became apparent that there was no suitable up-to-date text that was in print, in English, and dealt with the subject in a manner suitable for the advanced undergraduates and interested professionals. This represented a major gap since permafrost areas cover almost a quarter of the land area of the earth. Oil, gas and minerals are common in these areas, which therefore represent major reserves that will become increasingly important as the supplies of these substances become exhausted elsewhere.

The coming of modern medicines to permafrost regions has resulted in greater longevity of the indigenous peoples, giving rise to a population explosion. In North America, snowmobiles and aeroplanes have enabled hunters to greatly increase their productivity, which has destroyed the precarious balance between traditional food supply and the human population trying to subsist on it. Either the local indigenous peoples must move to warmer climates or they must change they ways of living.

This book presents a comprehensive review of the nature of the permafrost environment and the problems it poses for modern development. The successful construction and operation of roads, airfields, buildings, pipelines, etc. is extremely expensive in this inhospitable environment. Environmental processes are discussed, together with their effects. A small climatic change of ±2°C would greatly alter the situation and create problems in areas where Man is currently coping successfully.

Management problems are described in a non-mathematical way, giving examples from all over the Arctic, including Canada, Alaska, Siberia, Northern

Scandinavia, Greenland and Iceland. Low-latitude, alpine permafrost in the Alps, Tien Shan, Tibet and the cordillera of North America is also discussed.

Special care has been taken to provide adequate references throughout, so that the interested reader can study a selected topic in more detail. Wherever possible, the references are to papers in English, although it was necessary to include key papers and monographs in other languages.

ACKNOWLEDGEMENTS

The author is indebted to the following colleagues who kindly read and criticised earlier drafts of certain chapters: Professor N. B. Aughenbaugh (University of Mississippi, formerly School of Mineral Engineering, University of Alaska), Professor F. A. Campbell (Department of Geology, University of Calgary), Dr. D. C. Esch (Alaska Department of Transportation and Public Facilities, Fairbanks), Drs. C. Graham and R. Pilkington (Gulf Canada, now PetroCanada, Calgary), J. A. Heginbottom (Terrain Sciences Division, Geological Survey of Canada), and Dr. G. H. Johnston (Building Research Division, National Research Council of Canada. Dr. R. O. van Everdingen (Environment Canada, N.H.R.I., Calgary) kindly read the whole draft. Their help is gratefully acknowledged.

Illustrations are extremely important in explaining the concepts in a book of this sort. Almost all the line drawings are either originals or are redrawn especially for the book by Mrs. M. Styk. The photographs were taken by the author. It is my pleasure to thank the following for their permission to publish diagrams or tables redrawn from their publications: Alyeska Pipeline Service Co. (Figs. 7.14; 7.15); Arctic (Figs. 2.3; 2.4; 2.5; 3.1); Canadian Journal of Earth Sciences (Figs. 2.13 and Table 2.5); Dome Petroleum Ltd. (Figs. 7.1; 7.7); Esso Resources Ltd. (Figs. 7.5; 7.6; 7.10); the Geological Survey of Canada (Figs. 3.3; 5.5; 8.4); Hemisphere Publishing Corporation (Fig. 7.9); National Academy Press (Figs. 4.4; 4.5; 4.6; 5.6; 5.7; 5.9; 8.8); National Academy of Sciences (Figs. 2.9; 4.1; 4.3); Parks Canada (Fig. 2.6); Reinhold Book Co. (Fig. 2.19) and World Oil (Fig. 7.12). Grateful thanks are also due to those authors who also gave permission, including Professor D.E.Pufahl

(Fig. 2.19), and Dr. K. R. Croasdale (Fig. 7.8).

This book is the result of the interest in the problem engendered by the late Roger J. E. Brown, whose tutelage and colleagueship cannot be overestimated. Finally, I must thank my wife Pamela R. Harris for her companionship and encouragement on many field trips under inclement conditions and for her efforts in producing the camera-ready typescript.

Plate I: View into the Shargin Well in Yakutsk, USSR, that was described by Academician A. F. Middendorf in 1848. The moisture condensed as ice on the walls is removed at regular intervals.

Plate II: Building Showing the Results of Thaw Settlement in Dawson City, Yukon Territory, Canada.

Chapter One

INTRODUCTION

Permafrost is defined as ground that remains below 0°C for more than two years (Muller, 1943; Washburn, 1979), regardless of other properties such as moisture content and lithology. Glaciers and ice caps are excluded. Permafrost covers about 22% of the land area of the northern hemisphere plus substantial areas of the Andes and Antarctica. It also includes large zones of subsea permafrost along the southern margins of the Arctic Ocean (Fig. 1.1).

In spite of the large areas of occurrence, these regions have defied development by man until very recently. This book explores the distribution and properties of the permafrost and examines the difficulties encountered in constructing foundations, roads, railways and airfields in these areas. This is followed by an examination of the problems encountered by the oil and gas industry, by miners and by farmers. It will be shown that permafrost has a tremendous influence on the potential use and methods of development of a region.

HISTORY OF PERMAFROST SCIENCE

The Beginning

Undoubtedly people living in or visiting polar regions soon become aware of the existence of perennially frozen ground. The Vikings named the most icy land they encountered in their voyages across the North Atlantic, "Greenland" and a nearby more hospitable island, "Iceland", presumably to confuse other adventurers. Likewise the people of East Central Asia had lifestyles well adapted to the ubiquitous permafrost, using icings and springs for water supply, and earth-covered winter houses. The sod-houses of Iceland kept the Icelandic

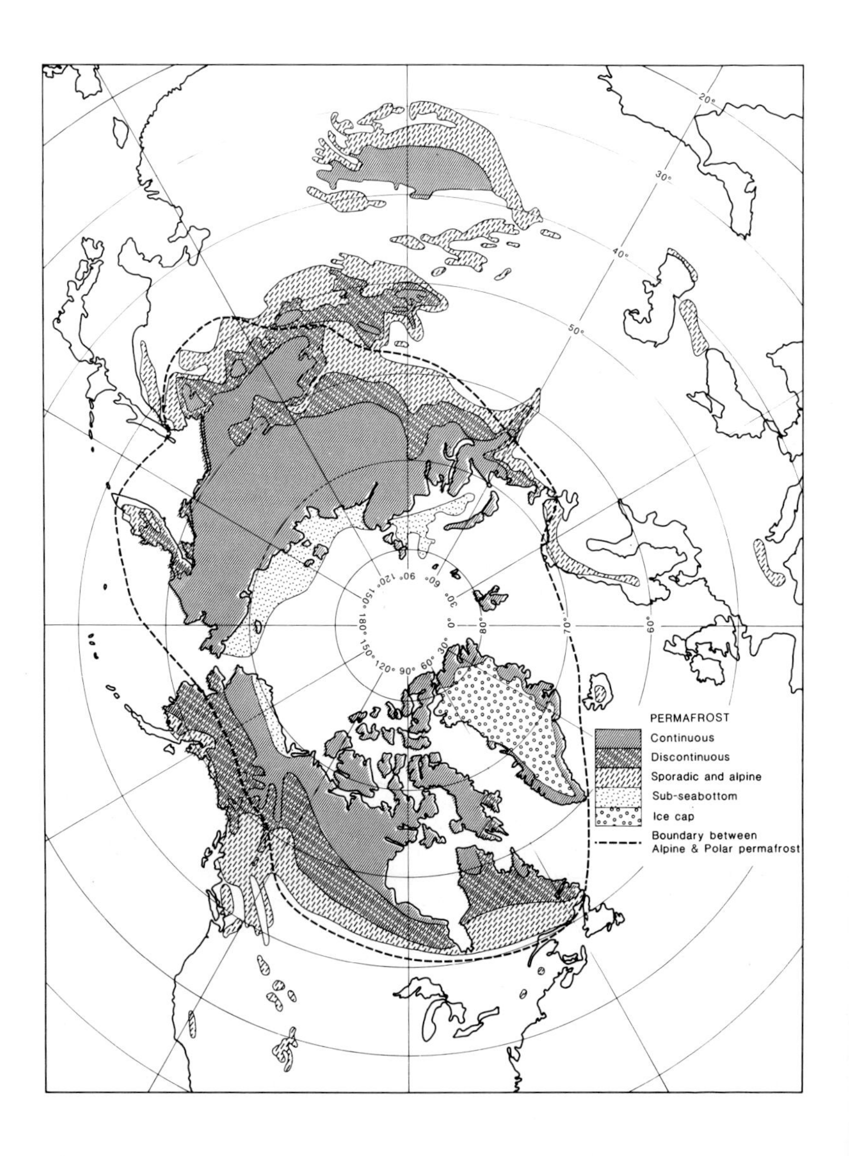

Figure 1.1: Distribution of Permafrost in the Northern Hemisphere (modified from Anon., 1985; Harris, 1983).

settlers warm, even though fuel was scarce, while Eskimos used blocks of compacted snow to build igloos. Both the Yakutians and the Eskimos were adapted to hunting and fishing, while the Icelandic peoples promoted grazing and fishing.

One of the first encounters with permafrost by English-speaking explorers was in the gold-mining attempts by Martin Frobisher on Baffin Island in the 1570s (Kenyan, 1975). Permafrost in North America was also discussed by Richardson (1839, 1851) and by Middleton (1743). The early descriptions were either the results of expeditions or of visits to the trading posts belonging to "The Governor and Company of Adventurers trading into Hudson's Bay" (now known as the "Hudson's Bay Company"). The Company was incorporated on May 2nd, 1670 and six of the eighteen adventurers named in the Charter of the Company were, or became, Fellows of the Royal Society. Among the earliest descriptions were those of Isham (1749) and Robson (1752), based on stays at Fort Prince of Wales (now called Churchill).

By keeping buildings simple and by building them on wooden sills or on stilts, there were very few problems in the early settlements. In fact it was not until the late 1930s that "The Company" decided to provide its post managers with improved living conditions. The first poured concrete basement and standard furnace were installed in a new mining settlement called Yellowknife on the northern shore of Great Slave Lake (Legget, 1965). The manager was the envy of the north because the furnace worked so well. Towards the end of the first winter, cracks appeared near the front steps. Then it was found that the house was settling. Remedial efforts proved ineffective and after 45 cm of settlement, the furnace was removed and the basement filled in. Heat from the furnace had passed through the basement wall and was melting the ice in the adjacent icy permafrost.

Observant travellers elsewhere also noted evidence for perennially frozen ground and some, such as Charles Darwin (1832), correctly deduced its cause. Darwin noted its presence in the South Shetland Islands. Recognition of perennially frozen ground as a problem was such that the Royal Geographical Society had an active committee on the "Depth of Permanently Frozen Soil in the Polar Regions, its Geographical Limits, and Relations to the Present Poles of Greatest Cold" (Lefroy, 1889).

Introduction

Russia expanded its territory from a limited zone around Moscow in 1490 by a series of military actions against the Mongols, so that, by 1692, it controlled the area north of Okhotsk, Irkutsk, Tomsk and the Caspian Sea (see Fullard & Treharne, 1972). Permafrost was mentioned in the Siberian military reports of Glebov and Golovin in 1642 (Tystovich, 1966). Initially, the Muscovites used the old ways of the indigenous peoples, e.g., the water supply consisted of springs which formed icings every year. As in North America, a trading company called the "Russian-Alaskan Trading Company" was incorporated to carry on trade in the new empire.

In the early 1800s, Shargin, the new governor of the Company, moved to Yakutsk and decided to try to obtain water from a well. After sinking a shaft 107 m into the ground, he ran out of money. All he found was very cold ground which grew ice crystals on its surface. Shargin sent a description of the well to the Russian National Academy of Sciences, asking for funds for further study. The scientists were surprised and sent Academician A. F. von Middendorf to look at the well in 1836. He measured the temperatures in the well and in other prospecting holes, and, although the well was deepened, the base of the perennially frozen ground was never reached (Middendorf, 1848, in Tsytovich, 1966).

From this study has evolved the modern Russian permafrost research, including the first known map of permafrost in Siberia by G. Vil'd in 1882 (Nikiforoff, 1928; Baranov, 1959). It was based on the assumed position of the mean annual isotherm of -2°C, as mapped using the limited climatic data available for the region at that time. This put the boundary too far north in Siberia and too far south in Europe. In 1895, I. V. Mushketov with V. A. Obruchev and others wrote the first "Instructions for Investigation of Frozen Soils in Siberia". Further publications followed with N. S. Bogdarov and A. N. Lvov describing the experiences of the engineers building the Trans-Siberian Railroad and searching for water supplies. Sumgin (1927) published the first textbook entitled "Permafrost Within the U.S.S.R.", which established the study of permafrost as an independent branch of science.

Development of Research Institutes in the USSR, Japan, the U.S.A., France and China

Since 1930, the USSR Academy of Sciences has been mainly responsible for systematic research in that country. It was necessitated by the decision after the Bolshevik revolution to systematically develop Siberia. Accordingly, the Permafrost Institute was developed at Yakutsk, together with other field stations where necessary, e.g., at Chernosevsky and Alma Alta.

Japan has negligible permafrost on its islands, but, one month prior to the attack on Pearl Harbour, it established the Institute of Low Temperature Science at Hokkaido University, Sapporo. This continues today, studying the distribution and properties of permafrost in polar areas.

Until Pearl Harbour, there was little systematic study of permafrost in North America. During the gold rushes, the prospectors had to develop new techniques of mining (see Chapter 8), while development of the oil resources near Norman Wells in an area of icy silts and clays was a matter of trial and error. Gradually, problems due to development of thermkarst were overcome but were not described until later (Legget, 1941). Similarly, the problems encountered in building the Hudson's Bay Railroad were not described for forty years (Charles, 1959).

After Pearl Harbour, there was a flurry of activity. Muller (1943) summarized the available knowledge on permafrost for the U.S. Army, including some from Soviet sources. The U.S. Corps of Engineers and the Canadian Army built the Alaska Highway from Dawson Creek, British Columbia, through the Yukon Territory to Fairbanks, Alaska in just over a year. It was supported by a chain of airfields and by a network of weather stations in the Arctic. Roads and pipelines were built across mountain ranges, e.g., the pipeline from Haines, Alaska, via Haines Junction, Yukon Territory, to Fairbanks, the Skagway-Whitehorse pipeline and the Canol pipeline from Norman Wells, Northwest Territories, to Whitehorse, Yukon Territory. In 1946, the U.S. Navy established its Arctic Research Laboratory at Point Barrow, while the University of Alaska, Fairbanks, had become the main centre for U.S. research. This work is continued today by the Cold Regions Research and Engineering Laboratory (CRREL) of the United States Army Corps of Engineers at Hanover, New Hampshire.

France has a permafrost institute called the Centre de Géomorphologie (CNRS) at Caen. It developed as an outgrowth of the work in Antarctica by French scientists but has specialized in the laboratory study of the effects of freezing and thawing on rocks.

China has some areas of alpine permafrost, and permafrost research is carried out as part of the work of the Research Institute of Glaciology, Cryopedology and Desert Research, Academia Sinica, at Lanchow. It carried out only limited work until the end of the Mao regime. With the acquisition of Tibet, China had to develop roads across the new region, which included extensive permafrost areas. The new regime is concerned with development of these areas, so permafrost research has greatly expanded (Jian & Xianggong, 1982).

PERMAFROST RESEARCH ELSEWHERE

In other countries, permafrost research is carried out by individuals and organisations that have developed a special interest in this subject. The only country with extensive areas of permafrost that operates this way is Canada. The Division of Building Research of the National Research Council of Canada was established in 1947 and, by 1950, it had begun to make a general survey of northern building. The starting point was a study of existing buildings along the Mackenzie valley from Hay River to the Arctic Ocean (Pihlainen, 1951). Studies were also made of the use of aerial photographs to recognize specific terrain conditions (Woods & Legget, 1960). Then a small research station was established at Norman Wells (Anon., 1956). The Division continued to lead Canadian research under the leadership of R. J. E. Brown, L. W. Gold, G. H. Johnston and R. F. Legget, until the establishment of the Canadian National Permafrost Committee in 1984. The Division was involved in choosing the townsites for Inuvik (Pihlainen, 1962) and Thompson (Johnston et al., 1963), and in studies of the ecology and distribution of permafrost.

The Division also established the Permafrost Subcommittee of the Associate Committee for Geotechnical Research. This Subcommittee organised regular meetings of the leading research workers on permafrost in Canada. It also organised the later Canadian Permafrost Conferences, the Third International Permafrost Conference in 1978 and the Secretariat for the International Permafrost Society. Many other organisations were also involved, including the Hydrology Division of Environment Canada, the Earth Physics Branch of the Department of Energy, Mines and Resources, and the Terrain Sciences Division of the Geological Survey of Canada (who are responsible for mapping permafrost). With the formation of the International Permafrost Society, the leadership may be changing to the National Permafrost Committee.

PUBLICATIONS ON PERMAFROST RESEARCH

These are scattered through the journals of several disciplines, as will be seen from the references at the back of this book. The CRREL series of publications is one of the largest series in English, while those of the Division of Building Research of the National Research Council of Canada also contain many useful papers. The best collections of papers in English dealing with all aspects of permafrost research will be found in the Proceedings of the four International Conferences on Permafrost and of the four Canadian Permafrost Conferences. Being a young science, there are at present no journals dealing with pure and applied research on the subject.

To aid the reader who wishes to obtain more details on specific topics, this book contains either a full bibliography on the topics or the references to the key summary papers. The best public library collections of papers on permafrost in English are in the libraries of the Arctic Institute of North America, housed at the University of Calgary, Canada and the Scott Polar Research Institute at Cambridge University, England. The World Data Center for Glaciology at Boulder, Colorado produced a reasonably complete bibliography of the permafrost literature between 1978 and 1982 (Brennan, 1983). The papers in the Canadian and International Permafrost Proceedings between 1958 and 1983 are indexed by Heginbottom and Sinclair (1985).

The Arctic Institute of North America maintains a multidisciplinary bibliographic database entitled the Arctic Science and Technology Information System (ASTIS). It is available to the public for on-line access and also produces a number of printed products including the bimonthly ASTIS Current Awareness Bulletin, in which "Soils and Permafrost" is one of the key subject divisions. Harris et al. (1986) provide a glossary of permafrost terms.

Plate III: Split Pingo, Tuktoyaktuk Peninsula, Northwest Territories. When a former lake drained and then froze, the soil water gradually changed to
Plate III: Split Pingo, Tuktoyaktuk Peninsula, Northwest Territories. When a former lake drained and then froze, the soil water generally changed to ice, thus expanding. This developed pressures in the unfrozen ground which exceeded the overburden pressure so that the excess water pushed upwards and froze to form the icy core of the pingo.

Plate IV: Sorted Stone Polygons in a Dried-up Lake-Floor, Denali Highway, Alaska.

Chapter Two

PERMAFROST IDENTIFICATION, NATURE AND PROCESSES

Since permafrost is ground that remains below 0°C for more than two years, it is necessary to make repeated accurate temperature measurements in the ground to prove its existence. Alternatively, there must be strong circumstantial evidence. This chapter will explore these methods, then examine the thermal and moisture regimes present, and finally it will discuss the processes occurring in permafrost areas due to seasonal and longer-term changes in these regimes.

MEASUREMENT OF GROUND TEMPERATURE

1. Sensors

The five main types of sensors used in measuring ground temperatures are contrasted in Table 2.1. In all cases the sensor must be buried in the ground and left undisturbed to allow the thermal regime to return to normal.

Platinum resistance thermistors can be very accurate but are far too expensive for field use. They are mainly used for calibration of other sensors in the laboratory. Thermistors are tiny bead resistors which change resistance with temperature. They give excellent results if properly sealed in the cable, and have become the standard for accurate field work since about 1975. They require an accurate resistance bridge (cost c. $1200) to read their resistance, which can then be converted to temperature. The response curve is not linear (Fig. 2.1).

Thermocouples consist of two wires of different metals or alloys whose ends are fused together. The junction generates a voltage which is dependent on the nature of the metals and the temperature. Copper and an alloy, constantan, are the cheapest

Table 2.1: Comparison of the Characteristics of Some Potential Ground-Temperature Sensors Using Hand-Held Measuring Devices

Type	Sensitivity @ 25°C	Accuracy for a Given Sensor	Accuracy Between Sensors	Cost	Comments
Platinum Resistance Thermistor	-0.4%/°C	±0.001°C	±0.01°C	$100 - $1000	The relatively large dissipation constant allows larger current flow for a given rise in temperature, giving better sensitivities than thermistors.
Thermistors	-4%/°C	±0.0025°C#	±0.01°C# to ±0.04°C	$ 15 - $ 400	Factory calibration. Has a smaller heat dissipation, thus requiring lower current flow to obtain maximum accuracy. Cost varies with stability and tolerance. Affected by moisture.
Thermo-couples	60μV/°C	±0.1°C	±0.5°C	$250 per 100 m	Low cost, not affected by moisture. No calibration, but use a stable reference junction and measurements of an ice-water bath between each determination for best results.
Diodes & Transistors	1.0mV/°C	≤±0.05°C	±0.05°C	$ 0.25 each	Needs individual calibration by user to obtain accuracies closer than 1°C. Affected by moisture.
Three Level Soil Temperature Recorders	±0.5°C	±0.5°C	±0.5°C	$1600.00	Needs individual calibration, preferably in field by comparison with a thermistor. Gives a continuous recording but can only be used at shallow depths (<3m). Charts must be changed monthly.

#If properly calibrated and installed.

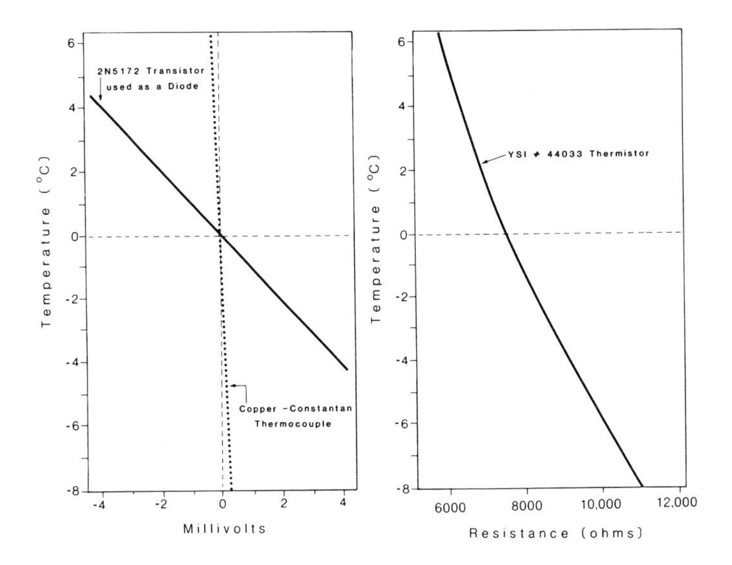

Figure 2.1: Comparison of Temperature and Output for a Thermistor, a Transistor used as a Diode and a Copper-Constantan Thermocouple that can be used in Measuring Ground Temperatures. The thermocouple has by far the poorest resolution.

and most convenient combination. The voltage changes sign at 0°C and the output is linear (Fig. 2.1). They require a very accurate potentiometer for measurement, plus regular calibration of sensor in an ice-water bath for the best results. Although thermocouples are cheap, the accuracy of the measurement is an order of magnitude worse than that of the thermistor. Hence they have been largely abandoned for ground temperature measurements.

Diodes and transistors are cheap, but need individual calibration by the user. They can be put into cables and sealed and provide results with an accuracy intermediate between those of the thermistor and the thermocouple. A meter can be built for about $50 to read them but the linearity of the response of voltage to temperature depends on the sensor that is used. The 2N5172 used as a diode gives a suitably linear response (Figure 2.1).

Perhaps the most useful, if limited, method is the three-level soil temperature recorder. One make, the Lambrecht, will operate at extremely low temperatures (down to -65°C) and provides a continuous recording of the temperature around the sensor. If one sensor is placed in an inverted white-coloured metal can, it can be used for measuring air temperature successfully on the tundra under conditions where the conventional Stevenson screen fills with snow. The other two sensors can be used for simultaneous continuous measurements of near-surface ground temperatures. The charts need to be changed and the clocks wound once a month. The data are accurate to ±0.5°C provided that a thermistor or transistor is used to establish a calibration curve in the field.

2. Drillholes

The sensors are mounted so as to form a string along a cable which can be emplaced in a hole in the ground. Ideally, the hole is drilled using an auger without lubrication so as to minimize thermal disturbance. Alternatively, a hotwater drill or a diamond drill may be used but leaves a thermal disturbance that may persist for up to five years. The hole may be cased to make it possible to change sensor strings but it is essential to ensure that the ground is packed back so that meltwater does not run down the drillhole in spring. A concrete collar may be necessary.

INDIRECT METHODS OF IDENTIFYING PERMAFROST

Where long-term measurements of ground temperatures are unavailable, other techniques may be employed, e.g. landforms, vegetation, climatic data, profile descriptions of drillholes and pits, and geophysical methods.

Landforms

These can be useful if the zonal permafrost landforms, i.e. those which are only found in permafrost regions, can be identified. Figure 2.2 shows the relationship of these to the permafrost zones. Unfortunately, in better drained or arid areas, conditions are not suitable for development of the diagnostic landforms. Thus, the distribution of permafrost in western Yukon Territory was found to be twice as extensive as previously believed (which was based largely on landforms) when the exploratory work was carried out

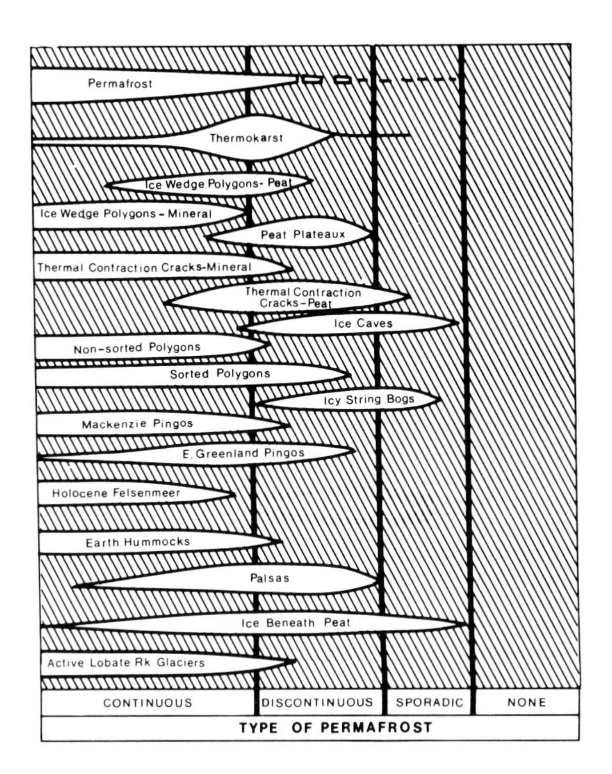

Figure 2.2: Relationship Between Zonal Permafrost Landforms and Permafrost Zones (after Harris, 1982b, p. 56). Reproduced with the permission of the National Research Council of Canada.

for the proposed Dempster lateral gas pipeline (Harris, 1983), and this has necessitated a major change in the permafrost map of Canada.

Vegetation

Certain types of vegetation, e.g. low herbaceous tundra, are always associated with permafrost. This is particularly true for plant associations such as *Dryas* which indicate negligible winter snow cover. However, other types of vegetation, e.g. shrubby willows and black spruce which are common in the subarctic, may not be confined to permafrost. Occasionally permafrost occurs under deciduous forest or under other species of conifers. Thus considerable experience with local conditions is essential if interpretation of the distribution of vegetation is to be used in locating the outermost boundaries of the permafrost zones.

Photointerpretation

This has been evaluated in many studies (Mollard, 1961; Mollard ¢ Pihlainen, 1966; Protas'eva, 1959, 1961, 1967; Frost et al., 1966; Hussey, 1962; Fletcher, 1964; Korpijaakko & Radforth, 1966; Pressman, 1963; Ray, 1960; Sager, 1951; Stoekler, 1948; U.S. Army, 1950) but produces considerable imprecision (Brown, 1974; Brown & Péwé, 1973). This is due to the dependence on recognition and interpretation of zonal landforms and vegetation patterns, discussed above. Wrong scale, poor photography, photography taken at the wrong time of year, poor lighting and inadequate ground-truthing compound the problem.

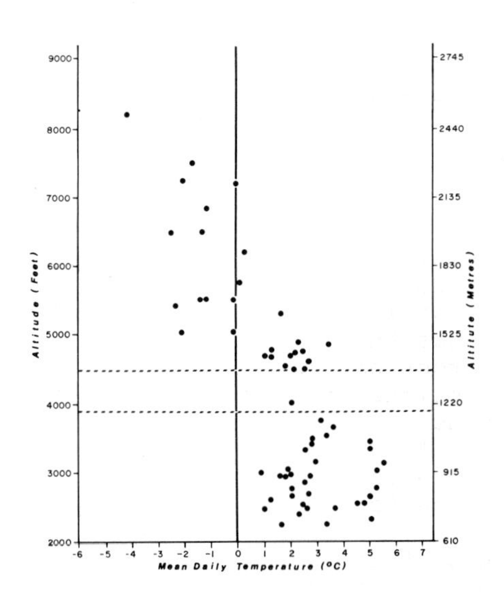

Figure 2.3: Mean Annual Air Temperature Versus Altitude for all the Class A Weather Stations Between 50° and 52°N in Alberta. The boundaries of the Parkland vegetation zone are indicated by the dashed lines around 1210 and 1350 m altitude (Harris, 1981a).

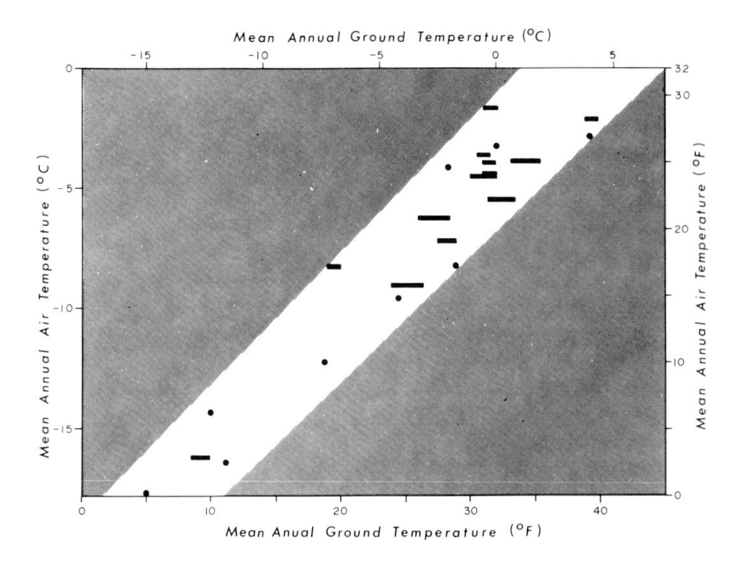

Figure 2.4: Mean Annual Air Temperature Versus Ground Temperature at the Top of the Zone of Zero Annual Amplitude (Harris, 1981a).

Climatic Data

The original permafrost maps of Russia in 1882, the Northern Hemisphere in 1928 (G. Vil'd; Nikiforoff, 1928) and for Canada (Brown, 1967) were based on climatic data. In each case, the authors had to use the available air temperature data, estimate what was the most likely correlation between permafrost boundaries and air temperature, and then extrapolate the results. In Russia, the permafrost boundary was plotted too far north in Siberia and too far south in Europe, while in western Canada, extensive permafrost was predicted too low along the Rocky Mountains (Harris & Brown, 1978, 1982).

The causes of the problem include local variability of mean annual air temperature (MAAT) with altitude in a given area (Fig. 2.3), lack of a precise correlation (Fig. 2.4) between MAAT and mean annual surface temperature (MAST) and considerable variability in the MAAT at the permafrost boundaries. At least part of the problem lies in the variability of winter snow cover which has been shown to act as an insulator (Brown, 1966b; Harris & Brown, 1978; Nicholson, 1978a; Mackay, 1978).

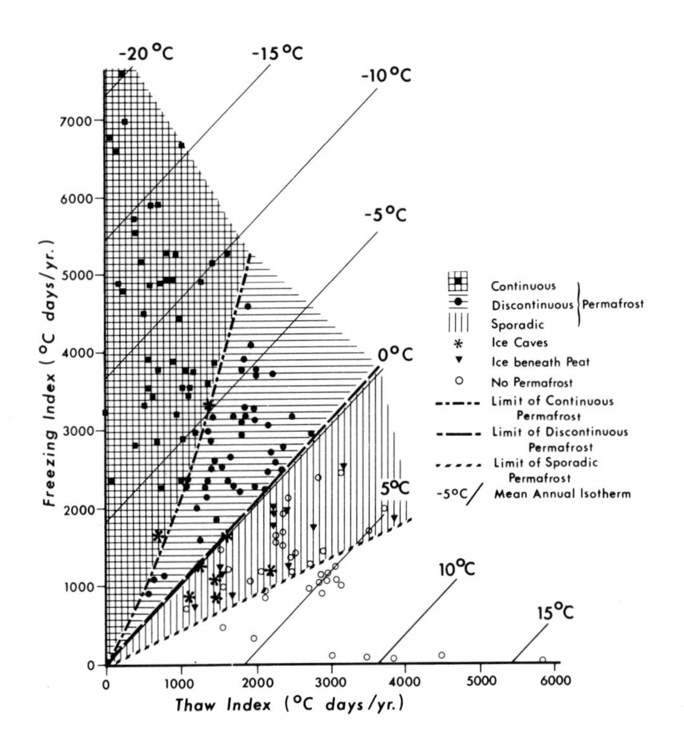

Figure 2.5: Relationship Between Occurrence of Permafrost Zone and the Plot of Freezing Index Versus Thawing Index for Stations in Canada, Norway, Spitsbergen, Iceland and the Mongolian People's Republic (after Harris, 1981a).

Permafrost distribution is reasonably predictable using climatic data if mean winter snow cover and mean daily air temperatures are available. The positive mean daily air temperatures for the year are added together to provide the total annual thawing index (Harris et al., 1986), while the sum of the negative temperatures for the year yield the total annual freezing index. For stations with a mean winter snow cover of under 50 cm, there is a clear relationship between permafrost zonation and freezing and thawing indices (Fig. 2.5). This method produces a reasonably satisfactory correlation of permafrost distribution in the western Yukon Territory with the data collected independently from over 100 ground temperature strings (Harris, 1983a). Without allowing for deep winter snow covers, the system predicts more extensive permafrost than actually occurs (see Greenstein, 1983).

For engineering purposes, the design thawing index is the average air thawing index of the three warmest summers in the most recent 30 years of records. For newer weather stations, the data for the last 10 years may be used (G. H. Johnston, personal communication, 1985). Similarly, the design freezing index is based on the average seasonal air freezing index (calculated by the method of Huschke, 1959) of the three coldest winters in the same data set. These are used in designs to predict probable extremes of thickness of lake ice, thickness of active layer and depth of frost penetration, as well as the ground thermal regime.

BTS Measurements

Haeberli (1973) has used measurements of the bottom temperature of the winter snow cover (BTS) to determine whether permafrost is present in the Alps and also the thickness of the active layer (Haeberli and Patzelt, 1983). They are also reported to work in Colorado (Greenstein, 1983) and Scandinavia (King, 1983). Unfortunately, BTS measurements do not work in continental, high latitude environments, e.g. Figure 2.6 shows the snow and ground

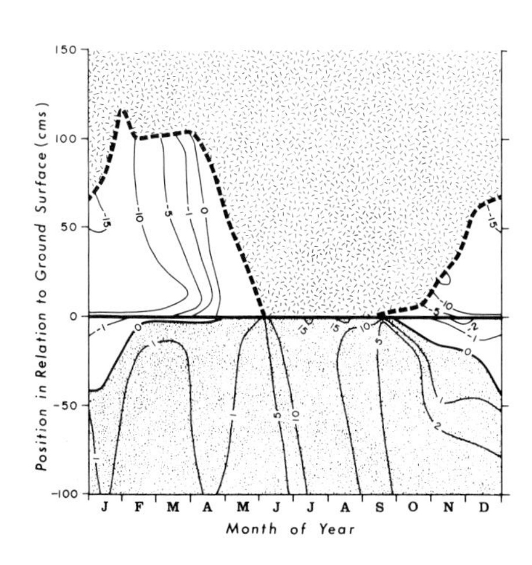

Figure 2.6: Geotherms for the Ground and Snow Cover at Site 2A, Vermilion Pass, British Columbia, measured once a week (Harris, 1976).

temperatures for site 2 in the Vermilion Pass, Kootenay National Park (latitude 51°N; elevation 1676 m) about 800 m below the lower limit of continuous permafrost (Harris, 1976). The BTS of -1°C would suggest that permafrost should be present by the interpretation of Haeberli.

Profile Descriptions of Drillholes and Pits

If moisture is present, a frozen layer should show up in drillholes and pits in the form of layers containing ice. This merely shows that the ground was frozen at the time of study but this can be an important first step in determining where permafrost may occur.

One method is to probe the ground with a long, pointed steel rod. It stops when it hits an icy layer and works well in peats, mucks, and unconsolidated sediments lacking stones. The alternative is to auger holes or dig test pits. Table 2.2 summarizes the ground-ice descriptive system used in North America (from Pihlainen & Johnston, 1963; Linell & Kaplar, 1966) and applied to all site investigations in potential permafrost areas. Where appreciable ice is encountered in late summer, permafrost will normally be present. However, the absence of obvious ice at a well-drained site may merely be a result of aridity, and temperature measurements will be needed to establish that permafrost is absent. Mackay (1972a) gives a thorough discussion of sources of ground ice in relation to the descriptive classification in Table 2.2.

Geophysical Methods

Geophysics has been used to map the distribution of permafrost on a routine basis for many years (Dement'ev, 1959). It is used to detect drastic changes in permafrost conditions between boreholes, to detect and delineate permafrost bodies in the discontinuous permafrost zone, to identify ice-rich layers *in situ*, to identify subsurface materials within the permafrost layer, and to determine the thickness of the active layer and of permafrost in areas of continuous permafrost (Scott & Mackay, 1977). They are the main method for mapping the probable distribution of offshore permafrost and detecting potential zones of gas hydrates.

The methods may be airborne or ground-surface, depending on the nature of the problem being examined. Airborne methods are most suitable for

Table 2.2: Ground Ice Description Used in North America (after Pihlainen & Johnston, 1963; Linell & Kaplar, 1966)

Group	Symbol	Description	Symbol	Comment
No ice visible	N	Poorly bonded or friable	Nf	Allow sample to warm in jar to estimate quantity of ice. Watch for ice coatings and reflective crystals.
		No excess ice	Nbn	
		Well-bonded excess ice	Nbe	
Visible ice < 2.5 cm thick	V	Individual ice crystals or inclusions	Vx	Describe ice phase using: Location / Size Orientation / Shape Thickness / Pattern and arrangement Length Spacing Hardness Structure Colour Volume of ice
		Ice coatings on particles	Vc	
		Random or irregularly-oriented ice	Vr	
		Stratified or distinctly-oriented ice	Vs	
Ice > 2.5 cm thick	ICE	Ice with soil inclusions	ICE + Soil	Describe as ICE, qualified as to hardness, colour, structure and inclusions.
		Ice without soil inclusions	ICE	

reconnaissance surveys of the thickness of taliks in areas of continuous permafrost and of the general distribution of discontinuous permafrost (Sellman et al., 1974). Ground surveys are used for detailed studies at specific sites or along linear transportation routes. However, relatively little success has been obtained in areas of thin permafrost (Fraser & Hoekstra, 1976).

The main methods used are resistivity, shallow seismic refraction and, more recently, electromagnetic dipole coupling (Henderson, 1980; G. H. Johnston, 1981). The resistivity method consists of passing a direct current through the ground from one electrode to another and measuring its resistance. The resistivity varies with soil type, water content, soil temperature and ice content. Provided that the geology is simple, measurements in different seasons quickly identify the thickness of the active layer and result in a reasonable estimate of the thickness of the permafrost layer. However, soluble salts greatly interfere with the results, since the current is carried in the liquid water films on the soil particles.

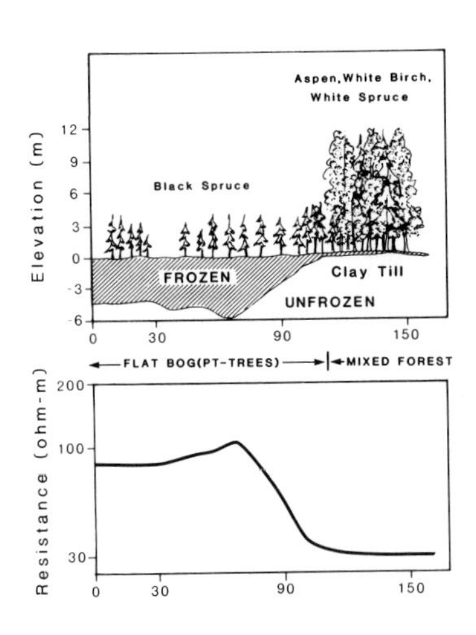

Figure 2.7: EM 31 Traverse, Mackenzie Valley Area (Henderson, 1980, p. 138). Reproduced with the permission of the National Research Council of Canada.

The other methods depend on transmission of seismic or electromagnetic waves through the material through the contacts between the grains. Ice tends to cement the particles together, thus improving efficiency of propagation of the waves (see Fig. 2.7 after Henderson, 1980, p. 138).

CHOICE OF TECHNIQUE

This depends on the specific problem; for building sites, drill holes, coring devices and thermistor strings are usually employed: for linear transportation routes, drill holes, thermistor strings and surface-geophysical methods between boreholes give the best results. Reconnaissance work involves the use of geological and topographic maps, aerial photographs and test sites with precise ground temperature data. This is the type of work that is being used by the Geological Survey of Canada in mapping permafrost distribution.

Generally, it is desirable to monitor ground temperatures for many years after a building or other structure is built to try to ensure that no problems will occur. Such monitoring installations would be invaluable in regions of changing climate.

For mapping offshore permafrost, the geophysical methods are the main source of information, augmented by borehole data and the distribution of submarine pingos/mud diapirs.

NATURE OF PERMAFROST

Permafrost in equilibrium with its environment consists of a layer of ground colder than 0°C near the surface of the earth (Fig. 2.8). It is overlain by a thin active layer which undergoes seasonal freezing and thawing; the temperature within the permafrost becomes warmer with depth. It is a result of a negative heat balance at the surface of the earth, where the mean annual surface temperature (MAST) is lower than 0°C.

Cutting through the permafrost may be unfrozen zones called 'taliks'. They are normally wet zones, often with flowing water. They tend to occur under streams, lakes or other "warm" places, and are subdivided into supra-permafrost, intra-permafrost and through taliks, depending on size and location. In North America, the unfrozen zone beneath the permafrost is sometimes incorrectly called a

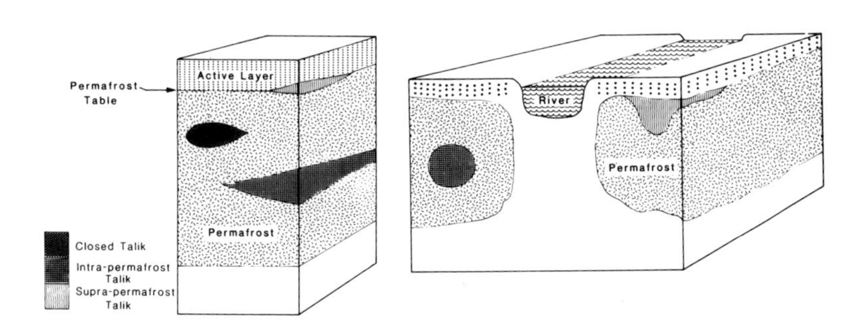

Figure 2.8: Permafrost Nomenclature

sub-permafrost talik. Where an unfrozen mass of sub zero °C saline water occurs, this is called a cryopeg.

The geothermal heat flow is believed to be fairly constant at about 1.6736×10^6 J m^{-2} yr^{-1}. The sun's heat has potentially 6,000 times greater effect at the surface of the ground (Judge, 1973a, p. 37), but this is greatly modified by factors such as snow cover, vegetation, thermal conductivity and soil latent heat. Thermal conductivity of rock varies considerably and sites with high conductivity will have a much thicker permafrost zone, other things being equal (Lachenbruch, 1970). Thus, permafrost is 650 m thick at Prudhoe Bay compared with 400 m thick at Barrow (Fig. 2.9), even though both have similar air temperature regimes today.

The thickness of the active layer varies from a few mm in the Arctic to over 20 m thickness at the outer boundary of alpine permafrost at lower latitudes (Harris & Brown, 1978, Fig. 7). Thirty cm to 2 m is a normal range in the Arctic and Subarctic.

Thermal Regimes in Equilibrium Permafrost

When the ground temperature measurements for a period of one or more years are plotted against depth, a trumpet-shaped curve is found that shows the effects of the seasonal variations in ground temperature (Fig. 2.10A). If the thermal conductivity of the material is constant, the geothermal gradient will be constant, and by projecting the slope upwards to the ground surface, it is possible to read off the MAST. A more usual way of comparing ground temperatures is to compare the ground temperatures at different sites using the depth of zero amplitude.

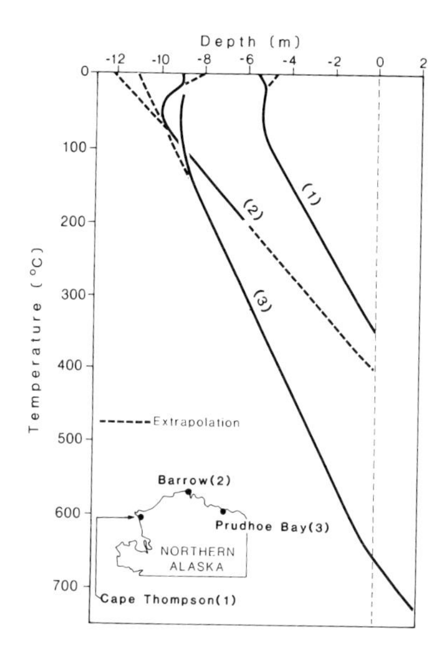

Figure 2.9: Geothermal Gradients in Boreholes in Arctic Alaska (after Gold & Lachenbruch, 1973, 10, Figure 1, and Lachenbruch & Marshall, 1969, 302, Figure 2)

This is arbitrarily defined as being the depth where temperature fluctuates by less than 0.1°C per year. In fact, fluctuations of lower magnitude can be traced to even greater depths by using thermistors, so Harris and Brown (1982) suggested calling it the depth of minimal amplitude.

Regardless of its name, the temperature at this level should be slightly higher than the MAST in equlibrium permafrost. Judge (1973b) reported that the temperature at the depth of minimal amplitude averaged 3.3°C higher than the mean annual air temperature for 38 sites in Canada, with a standard deviation of ±1.5°C (in other words, it is rather variable), thus agreeing with Brown (1967, 1970a).

Figure 2.11 shows the geotherms for site 2A at Plateau Mountain in south-western Alberta. It demonstrates that both the seasonal heating and cooling waves take time to penetrate into the ground, e.g. approximately 8 months for 15 m in this case. The rate depends on the thermal conductivity of the rock. This is a dry site and it will be seen that the active layer refreezes from below as well as from

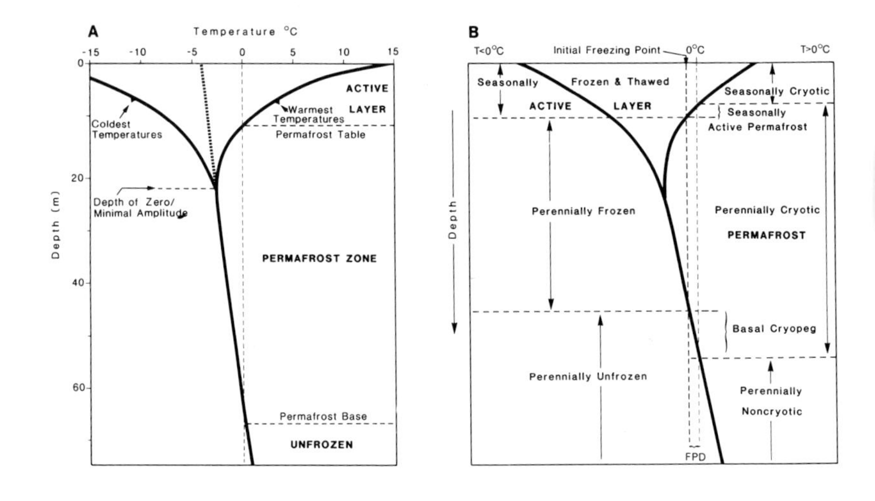

Figure 2.10: Typical Temperature Envelope for an Area with Permafrost (2.10B, after van Everdingen, 1985). FDP is Freezing-Point Depression.

above. This up-freezing varies with moisture content from year to year but can reach 3 m in some years at this site. This contrasts with the 2-13 cm of up-freezing first reported from the wet silty soils of the Mackenzie Delta (Mackay, 1973b). Upfreezing, first demonstrated by the U.S. Corps of Engineers in Washburn (1956), is very important in that the direction of the movement of the freezing front is opposite in the upper and lower portions of the active layer. This is commonly ignored in many theories concerning active layer processes.

Thermal Regimes in Disequilibrium Permafrost

A close look at Fig. 2.9 shows that the upper part of the temperature profile shows some surprising oscillations. These are interpreted as being evidence of climatic change (Gold & Lachenbruch, 1973; Lachenbruch & Marshall, 1969).

The two possible changes that can modify an equilibrium temperature profile are climatic warming (Fig. 2.12A) or cooling (Fig. 2.11B). This can also be achieved by a change in thickness of the mean winter snow cover. This changes the amount of heat flow from above and brings about a progressive change in ground temperature regime starting at the surface (Lachenbruch, 1968). The rates of such

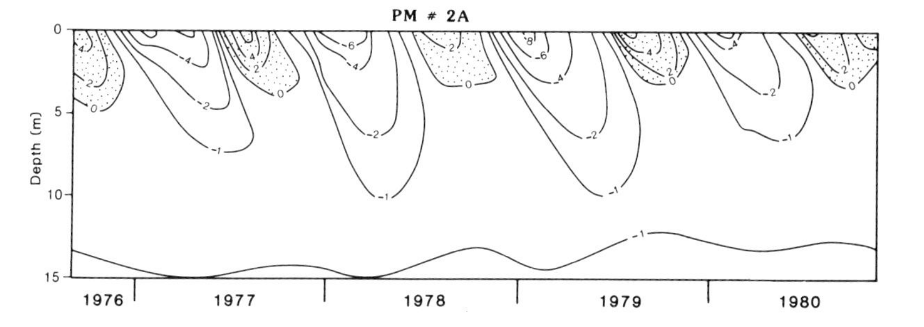

Figure 2.11: Geotherms for Site 2A at Plateau Mountain, southwest Alberta

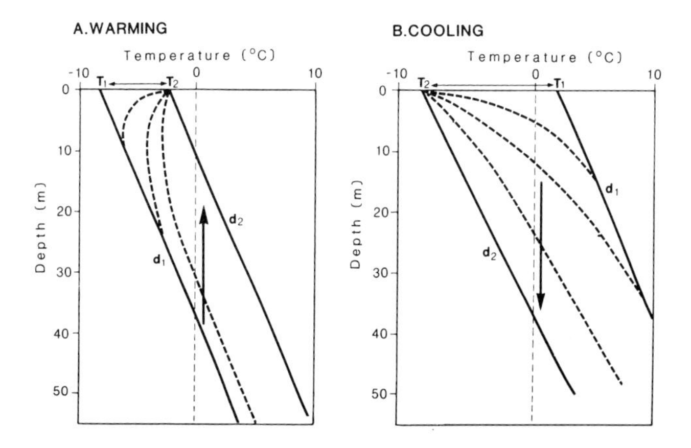

Figure 2.12: Theoretical Effects of Climatic Change from Mean Ground Surface Temperature T_1 to T_2 on the Mean Ground Temperatures at Varying Depths with Time (after Lachenbruch, 1968). d_1 and d_2 refer to the initial and final equilibrium depths of the base of the permafrost table.

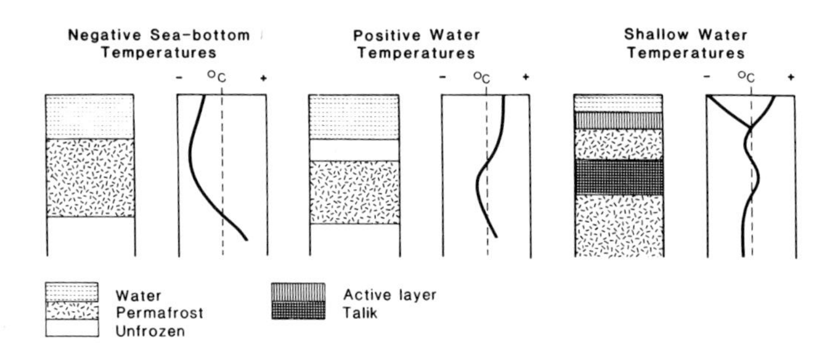

Figure 2.13: Thermal Conditions of Disequilibrium Beneath Various Types of Water Bodies (after Mackay, 1972b; Hollingshead et al., 1978). The shallow water differs from the other cases by becoming frozen to the bottom each winter. In part, reproduced with the permission of the National Research Council of Canada.

changes have been calculated by Terzaghi (1952). For a soil with 30% ice by weight, the rate of rise of the base of the permafrost zone would be 200 m per 10,000 years, i.e., 2 cm/year. However, a 2°C rise in MAST would thaw 15 m of the top of the permafrost table in 100 years.

The small amount of change in the permafrost base has been used by Balobayev (1978) to calculate the minimum mean annual air temperatures for the Wisconsinan, based on the thickness of the permafrost in the Wisconsinan terraces.

The curves in Figure 2.9 indicate that Barrow and Prudhoe Bay had been subjected to a former MAST of -12°C for a long period of time. This was followed by a warming of 4°C since about 1850 A.D., and then a warming of 1°C since about 1940-1950. Similar disequilibrium conditions have been reported from Spitsbergen, with a warm period between 1920 and 1960 A.D. (Liestøl, 1977).

Disequilibrium temperature profiles have also been reported from beneath water bodies (Fig. 2.13, after Mackay, 1972b; Hollingshead et al., 1978). The negative sea-bottom temperatures are typical of shallow parts of the Arctic Ocean, e.g. the Beaufort and Laptev Seas, while the positive water temperatures are encountered along the major rivers flowing into the Arctic Ocean. The shallow water temperatures are encountered in oxbow lakes and other shallow lakes.

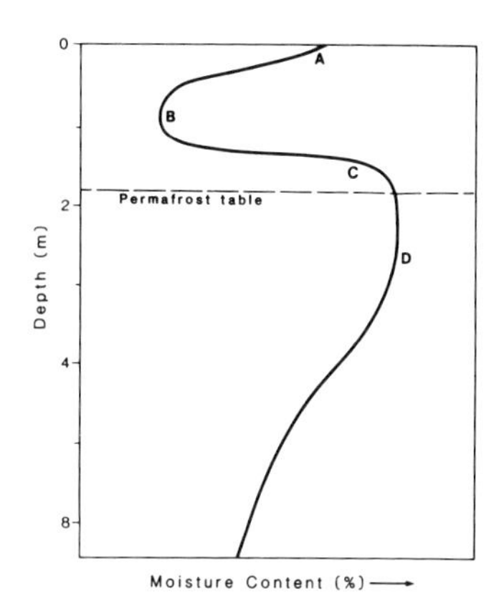

Figure 2.14: Typical Profile of Total Moisture Content Through the Upper Layers of Permafrost After Freeze-Back in Winter. For an explanation of the origin of the parts marked by letters, see the text.

Moisture Regimes in Permafrost Areas

Permafrost is a temperature condition, but one of the bigger differences between temperature profiles over permafrost and those in non-permafrost regions lies in the moisture regime. Figure 2.14 shows a typical temperature profile through a soil over permafrost at the end of freeze-back. The maximum, A, is partly due to snowmelt infiltration into the surface horizons of the soil, prior to the ground freezing. It is augmented by water moving to the freezing plane from below during freezing (as will be discussed below). Maximum C corresponds to the frozen perched water table at the base of the active layer from the previous summer season, also augmented by moisture moving from the middle of the active layer to the freezing plane that advanced from below. Both maxima (A & C) are characterized by ice lenses (Mackay, 1981a). Minimum, B, corresponds to the zone of maximum depletion of water during freeze-up.

The upper part of the permafrost (D) usually contains the largest quantities of ice, often as massive icy beds (Mackay, 1971a, 1971b; Brown & Sellman, 1973). This appears to be true regardless of the form of the ice or the position of the profile on the landscape. Individual ice crystals tend to be 14-25 times larger than in seasonally frozen ground unless recrystallization has taken place after shearing (Tsytovich, 1975). Actual ice content may be 2000% by dry weight in peat, even in the absence of ice lenses (Kinosita, 1978), and amounts in excess of 1000% by dry weight are not unusual in the upper layers of permafrost developed in silts. Massive icy beds commonly contain less than 0.1% soil by weight (Mackay & Black, 1973) and the base of such layers usually lies within 48 m of the surface of the permafrost table (Mackay, 1973a).

The seasonal and longer-term dynamics of the moisture regime are now under intensive study but it is already apparent that moisture can move from the unfrozen active layer into the frozen active layer and even into the top of the permafrost (Cheng, 1982; Lagov & Parmuzina, 1978; Mackay, 1983; Parmuzina, 1977, 1978, 1979; Pikulevich, 1963; Tystovich, 1975; Vtyurina, 1974; Williams, 1984; Wright, 1981; Zhigarev, 1967). This is usually ascribed to migration of pore-water in the direction of the coldest temperatures. However, substantial increases in the moisture content of the active layer can be demonstrated on top of Plateau Mountain, southwest Alberta (Fig. 2.15) during spring when air temperatures do not rise to 0°C but the snow pack is warmer than the underlying ground. The soil is a loamy blockfield. When the snow-pack becomes isothermal at the end of May, percolating water freezes on the surface of the cold ground, forming a layer of ice up to 5 cm thick, similar to that described by Woo and Heron (1981) from the high Arctic. This essentially stops moisture movement into the still frozen active layer and the melting snow evaporates under the influence of the strong mountain winds.

After the snowpack has gone, thawing of the icy active layer proceeds rapidly, producing an extremely wet zone above the frozen active layer. Mackay (1983) describes the evidence for moisture transfer from the thawed zone into the still frozen active layer and even the upper portion of the permafrost by movement of unfrozen pore water under thermal and/or hydraulic gradients. The net result is enrichment of the surface of the permafrost zone with ice, and it also demonstrates that it is difficult to prevent moisture build-up in the upper layers of permafrost.

Figure 2.16 shows the perched water table at site A at Summit Lake, northern British Columbia in July 1983, measured by neutron probe. About 1.5 cm

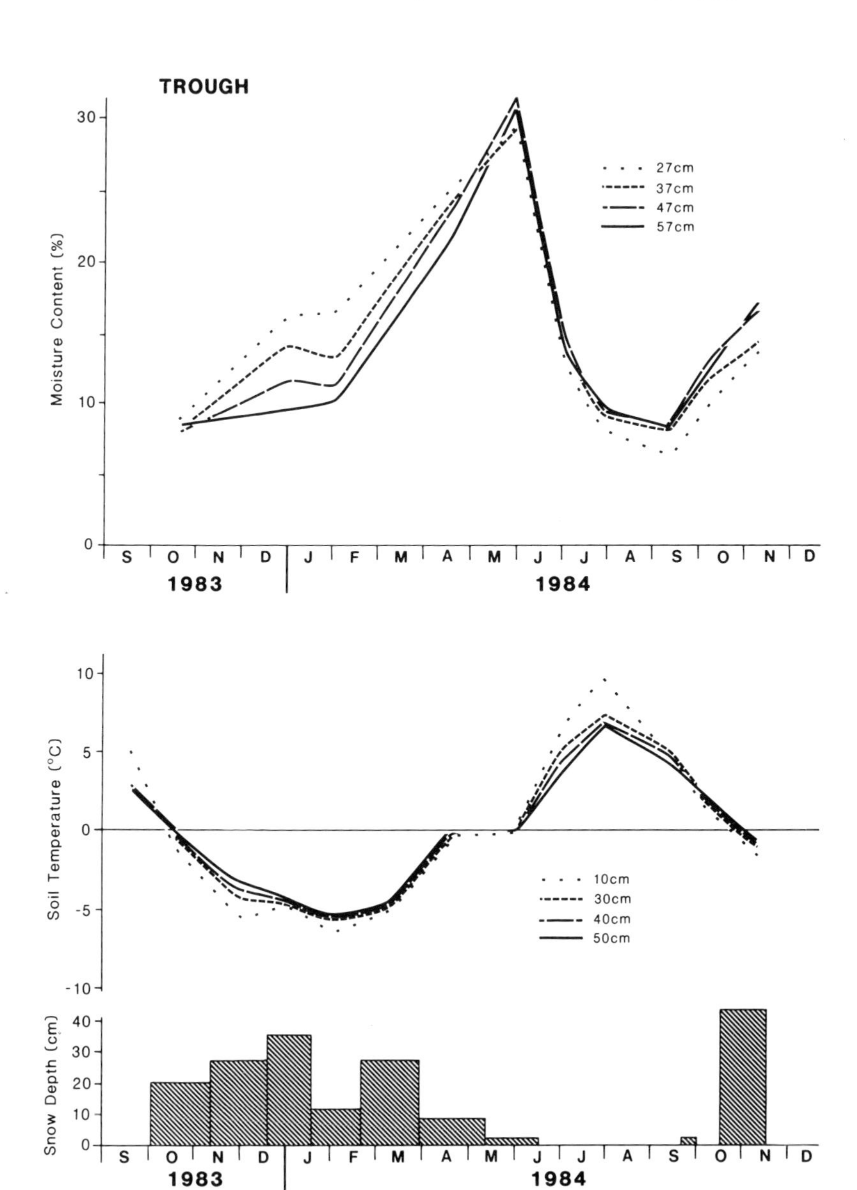

Figure 2.15: Moisture Input into Frozen Ground at Plateau Mountain, Alberta (Site 1) During the Winter of 1983-1984, as Measured by Neutron Probe. Moisture content expressed on a weight basis.

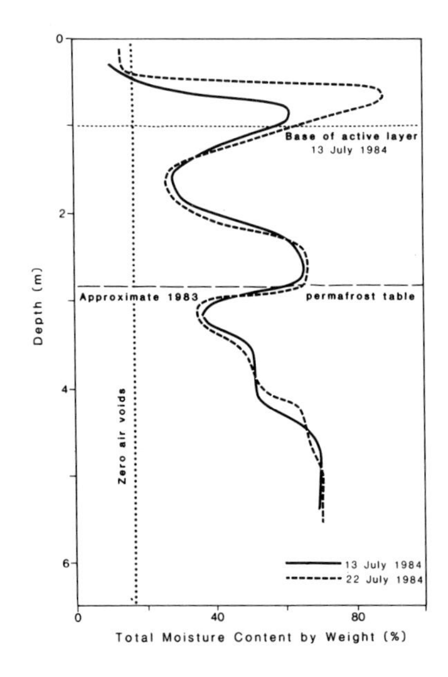

Figure 2.16: Perched Water Table at the Base of the Thawed Layer at Site A, Summit Lake, British Columbia. Moisture is accumulating in the permafrost table following thawing during drilling of the hole.

of rain fell between the two observations and it will be seen that the rain augmented the meltwater from ice in the active layer. Note also the redistribution of moisture taking place at depth. This perched water table, containing water from all available surface sources (melting snow, melting ice and rain) is of considerable importance since it provides a continuous water supply to plants during the limited growing season, and means that there is always a zone of limited cohesion above the frozen ground in all except the most arid areas (see Woo, 1976; Woo & Steer, 1982). The perched water table also has a tremendous effect on slope stability and often results in active-layer failures which can develop into retrogressive thaw slumps (Hughes, 1972; Mackay, 1966; McRoberts & Morgenstern, 1974a, 1974b; Rampton & Mackay, 1971).

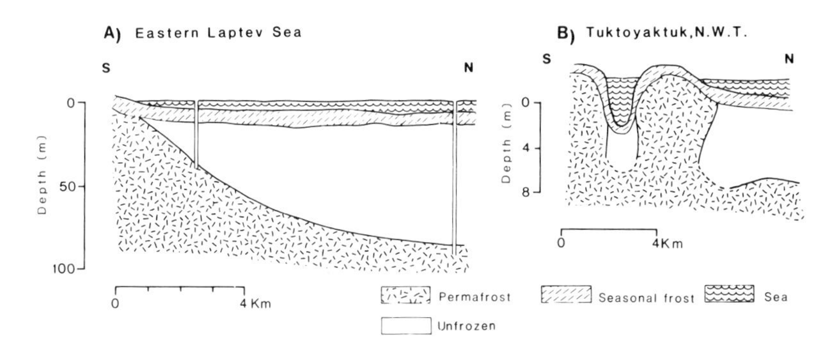

Figure 2.17: Offshore Permafrost in (A) the Eastern Laptev Sea (after Vigdorchik, 1977) and (B) Tuktoyaktuk, Northwest Territories (modified from Shah, 1978 and reproduced with permission of the National Research Council of Canada).

OFFSHORE PERMAFROST

This was first reported in Spitsbergen by Werenskiold (1922) and is common around the margins of the Arctic Basin (Hopkins & Sellman, 1984). The approximate boundaries of known subsea permafrost are shown in Figures 7.5 and 7.6. The evidence for thick ice-bonded sediments in the cliffs of the coast and under the bed of the Beaufort Sea is discussed in Morack et al. (1983), Neave and Sellmann (1983), Blasco (1984), Hunter (1984), and Sellman and Hopkins (1984).

The actual occurrence of permafrost varies enormously from place to place and the determination of its exact position and cause is quite a challenge. Two cross-sections of shores described by various authors are shown in Figure 2.17. The existence of permafrost at these sites was established by drill holes and thermistor strings but geophysics is used for most of the mapping. The geophysical interpretations must then be checked and refined by drilling.

The variability is partly due to water temperature, variations in salinity, type and rate of deposition of the sediments, coastal erosion and changes in sea level relative to the land (Werenskiold, 1953; Lachenbruch, 1957; Mackay, 1972a, 1972b; Jenness, 1949; Hunter et al., 1976; Judge, 1974). Most cases of offshore permafrost are probably relict and in disequilibrium with the present thermal environment. It is often quite thick (800 m, Fig. 7.2) and developed during colder conditions during the Quaternary period when the sea level was considerably lower than today and the shallow parts of the present sea floor were above sea level. If the seawater is warmer than 0°C, then the permafrost will be degrading from above and below, but if the seawater temperature is below 0°C, then it will be degrading from below (see Fig. 2.13). In the latter case, the thawing is solely due to geothermal heat and Mackay (1972b) has calculated that there has been too little time for complete degradation of the relict permafrost.

If the coast is fairly stable and the sea bottom temperatures are negative, equilibrium permafrost can develop in the sediments beneath the sea floor. Vigdorchik (1980) calculates that this will rarely exceed 200 m in thickness. The exact temperature at which ice-formation occurs will depend on the salinity of the sediments. Sands with non-saline pore-water will become ice-cemented just below 0°C, but freezing-point depression will occur in saline materials, while clays may be plastic.

Where fresh water comes to the surface of the sediments under artesian pressure in spring, submarine pingo-like features may develop (Shearer et al., 1971). Small conical mounds were detected on the shallow sea floor throughout much of the Arctic by the S.S. Manhattan. The temperature of the sea floor is between -0.5°C and -4.0°C, and the mounds vary from 300 m to 1000 m in diameter and rise to within 18 m of sea level, which is about the depth of the halocline. It seems unlikely that pingos formed on land could survive submergence, as they would be above the halocline in summer and so experience above-zero temperatures, yet more mounds are still being found. It remains to be established whether this is due to inadequate mapping or growth of new mounds but they are now thought to be mud diapirs (Poley, 1982). Whatever their origin, these mounds represent severe problems for tanker traffic.

Along shorelines that are changing position, permafrost is invariably in disequilibrium. Rapid deposition, e.g., on the Mackenzie Delta front, will result in buried permafrost with sediments containing aggrading permafrost. Where localized rapid erosion takes place, the permafrost will be exposed on the sea floor and in the beaches and cliffs, as at Tuktoyaktuk (Fig. 2.17B, modified from Shah, 1978).

These represent unstable thermal environments that need special care with design, if they must be used.

Inevitably, subsea permafrost creates a different set of geotechnical problems for the petroleum industry which are briefly discussed by Hayley (1984) and by Jahns (1984a).

THE FREEZING PROCESS

It is important to understand what happens during freezing. Frozen ground normally consists of at least four constituents, viz.: mineral and/or organic matter, air, ice and water. The liquid water is particularly importance since it can move about.

Water is held in soil or rock under conditions ranging from infilling of pores (gravitational water) to a thin molecular layer surrounding the individual solid particles. The proportion of the latter depends primarily on the surface area of the solid phases but may reach 30% of the dry weight of the soil in clays. Since the gravitational water is held under the influence of gravity, it is free to change phase at or near 0°C. Hosever, the adsorbed or interfacial water is bonded by strong electrical forces to the mineral particles. This is because water is a polar molecule (Fig. 2.18) with different surface charges at different positions of the triangular structure. Since the surfaces of minerals also have positive and negative charges at different positions on their surface, the water molecules tend to become aligned with about 40 films of molecules bonded to the mineral surface with positive charges against negative charges. The strongest bonding is in the inner layer and the bonding forces decrease outwards. After this, some water is held by capillary forces between adjacent grains, but these forces are intermediate between the bonding forces

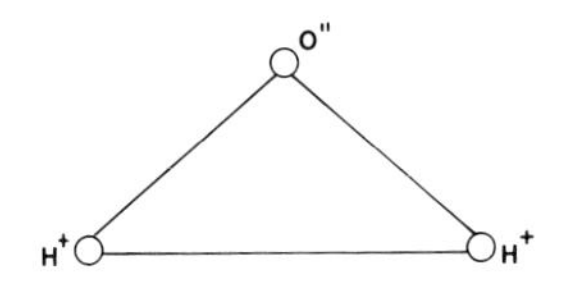

Figure 2.18: Structure of a Water Molecule, Resulting in Positive and Negative Sides Producing the Dipole Effect (after RIGCDR, 1981). Reproduced with the permission of the National Research Council of Canada.

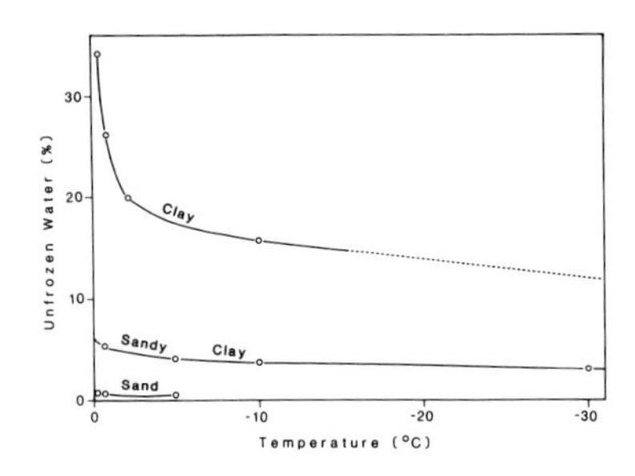

Figure 2.19: Variations in Liquid Water Content of Different Materials with Temperature (after Tsytovich, 1957).

and the gravitational forces. Since the quantity of adsorbed water depends on grain size, there are considerable differences in liquid water contents between soils of different textures (Fig. 2.19, after Tsytovich, 1957).

When the soil is cooled to or below 0°C, it will tend to cause the gravitational water to crystallize. The exact temperature at which this can occur depends on pressure, mineralogical composition, exchangeable cations and solute content. It is this variability in the exact temperature at which freezing commences that makes it necessary to define permafrost in terms of temperature.

Gravitational water may remain as a liquid to perhaps -3.0°C and then it commences freezing (Fig. 2.20). As it does so, it releases 3.347×10^8 J m^{-3} of latent heat of fusion. This causes the temperature to rise to about 0°C and remain there until the rate of heat extraction from the material can no longer be counteracted by the rate of release of latent heat. At this point, the temperature of the ground will start to drop. As it does so, the freezing forces will progressively exceed the capillary forces so that some of the more loosely held capillary water will also change into ice. However, the molecular bonding forces are so great that not all the adsorbed water will freeze, even at -78°C.

Salts depress the freezing point of gravitational water, particularly in saline soils, cryopegs and offshore permafrost. Pressure is also very important since at 100 atmospheres, ice melts at -0.9°C, while at 1500 atmospheres, the ice melts at -14.1°C. Within

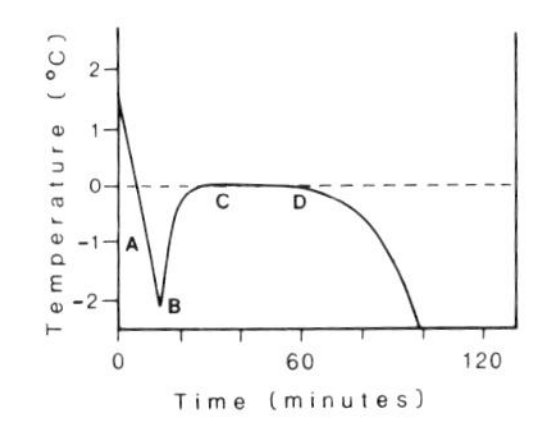

Figure 2.20: Ground Temperature Changes at the Onset of Freezing (modified from RIGCDR, 1981):
A Supercooled water.
B Commencement of crystallization of water phase.
C Latent heat of fusion released by freezing negates all the negative heat flow coming from above.
D Commencement of surface cooling exceeding latent heat of fusion, resulting in cooling and downward movement of the freezing front.
Reproduced with permission of the National Research Council of Canada.

the soil, the contact surfaces of the grains take all the pressure, so that even under a light load, the ice will melt on the contact surfaces. This probably partly explains the relative lack of massive icy beds and ice lenses at greater depths in permafrost. The amount of liquid water in the soil has a tremendous influence on freezing rate, since it only requires about 1.8 - 2.1 X 10^8 J m^{-3} to lower 1 g of soil through 1°C (i.e., 1/160th of the latent heat of fusion of ice). Berg (1984) discusses the currently available numerical models for heat and moisture transfer in frost susceptible soils.

Actual Freezing of Soil

Under natural conditions, ground freezing commences with the very top layer. Needle ice develops with the crystals growing at right angles to the freezing plane. This prevents supercooling of the underlying soil moisture. If the soil is fairly dry, supercooling of the water occurs until the forces of crystallization overcome the other forces. Then the temperature will rise back to 0°C as freezing occurs (see Fig. 2.19).

Once a freezing plane has been established, liquid water moves from below to the freezing plane, which tends to desiccate the underlying layers. The freezing plane stays in place until the latent heat produced by freezing of water is less than the heat loss to the surface. Then the freezing plane slowly

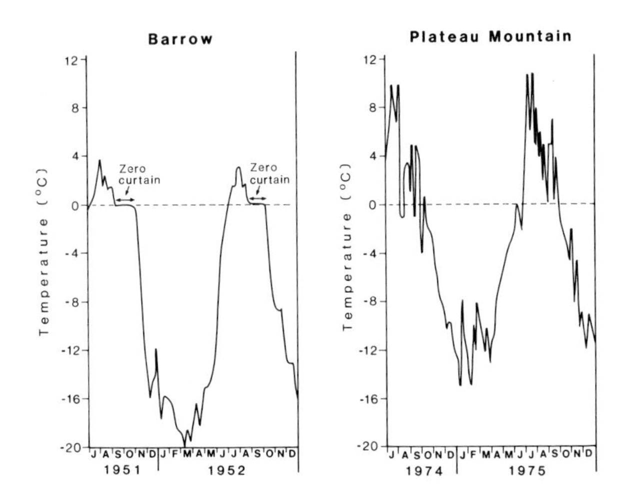

Figure 2.21: Temperature Distribution Versus Time for a Depth of 25 cm Near Barrow, Alaska (after Brewer, 1958, Fig. 4), and for a Depth of 20 cm at Site 2, Plateau Mountain (2,499 m elevation), Alberta. Note the presence of the "zero curtain" effect in the Fall in the wet soil at Barrow and its absence at the drier Alpine site.

descends. The period of constant temperature at 0°C at a given level in the soil may show up in ground temperature measurements; it is called the "zero curtain" effect (Fig. 2.21). It is most commonly observed in very wet silty soils.

The movement of water to the freezing plane is accomplished by electrostatic and osmotic forces (Scott, 1969), giving what is often called the suction potential or water potential of the freezing system (Anderson, 1971). The dipole effect of freezing water causes re-orientation of the adjacent liquid water molecules and the generation of electrical freezing potentials which may reach as high as 1200 mv in lacustrine clays at Illisarvik (Parameswaran & Mackay, 1983).

There is also usually some upfreezing from the cold permafrost table. This results in similar changes taking place at the base of the active layer, the importance of which is only just being realized. Gold (1985), Smith (1985) and Washburn (1985) discuss models of soil freezing and their limitations in field situations.

Frost Heaving

When water changes to ice, it increases 9% in volume. In soils with less than 90% of the pores occupied by water, and where no additional water comes from elsewhere, the ice will tend to cement the grains together. However, where more ice is formed, there will be heaving of the overlying layers in proportion to the excess ice (Taber, 1929, 1930a, 1930b). The energy for this is provided by the latent heat of fusion. Most soil freezing is essentially an open system, with the heaving resulting largely from the influx of water from outside. Direction of heave is parallel to the direction of crystal growth, which will be indicated by the elongation of inclusions of air bubbles (Gell, 1974). The amount of heave produced when the temperature of the soil drops just below 0°C and remains there for a long time may be similar to that produced by an extreme drop in temperature for a short time. However, prediction of the amount of heave in a given situation has proved very difficult (Anderson et al., 1984; Grechishchev, 1984; Chen Xiaobai, 1984).

Heaving of soils wetted with seawater is about half of nonsaline soils under otherwise identical conditions (Chamberlain, 1983). This is associated with greatly reduced ice lensing. Seawater causes the formation of a thick active layer with many growth sites for ice lenses, each with its own brine concentration. Unbonded brine-rich zones between ice lenses are identified as potential zones of low shear strength.

Thaw Settlement and Densification

The amount of thaw settlement following heaving in a closed system may exceed the amount of heaving by as much as 20% in the first few cycles (McRoberts & Nixon, 1975, p. 162). This is due to densification and results in overcompacted sediments. The effect is rather similar to consolidation of sediments during drying (Williams, 1967, pp. 1-10). It also causes changes in the thermal and geophysical properties of the soil.

Overcompacted sediments of this type have been described by Hollingshead et al. (1978) from the Mackenzie Delta, where they contrast quite markedly with the new sediments forming today. The soils have a dense, compact, brittle structure which parallels the ground surface. The compaction decreases with depth, i.e. these soils can be classified as a type of fragipan (see Grossman & Carlisle, 1969; Ruhe, 1983). Although fragipans have been interpreted as indicating permafrost conditions (Langohr & Van Vliet, 1981), they can be formed in other ways under other climatic regimes.

Ice Segregation

This has been studied extensively since the pioneer work of Taber (1929, 1930a, 1930b). Ice segregation is favoured by materials with grain sizes of 0.01 mm or less. However, if the material is too fine-grained, the permeability will be too low to permit the movement of sufficient quantities of liquid water. Thus silts are the soils in which ice segregation is most common.

Mackay (1971b) has summarized the current ideas on the origin of segregated ice. It is favoured by high pore-water pressure. Where the freezing plane remains stationary for long periods of time, water moves freely to it, ice crystals form and produce an ice lens parallel to the freezing front. Should the water supply be inadequate for the pore-water pressure to exceed the overburden pressure, the freezing plane tends to move downwards unevenly, so that tongues of ice extend downwards. The water in the rest of the soil freezes as pore ice. Fine-grained soils liable to ice segregation are said to be frost-susceptible.

Frost-Susceptibility of Soils

Unfortunately, there is no generally accepted criterion that characterizes a frost-stable material. The criterion most commonly used is grain size (R.C. Johnson et al., 1974). Minimum grain size, amount of fines and degree of mixing (the grading) are all important.

Casagrande (1932) stated that, given a sufficient water supply, one should expect considerable ice segregation in non-uniform soils containing more than 3% of grains finer than 0.02 mm diameter, and in very uniform (well graded) soils containing more than 10% finer than 0.02 mm. No ice segregation was observed by Casagande in soils containing less than 1% of grains finer than 0.02 mm, even when the water table and frost line were coincident.

This remains the best general system according to Linell and Kaplar (1959). The frost design soil classification system developed by the U.S. Corps of Engineers (Table 2.3) is based on it. The soils are listed in their general order of increasing susceptibility to frost heaving and/or weakening as a result of thawing. Thus the F1 sediments should

Table 2.3: U.S. Corps of Engineers Frost Design Soil Classification

Frost group	Soil type	Percentage finer than 0.02 mm, by weight	Typical soil types under Unified Soil Classification System
F1	Gravelly soils	3 to 10	GW, GP, GW-GM, GP-GM
F2	(a) Gravelly soils	10 to 20	GM, GW-GM, GP-GM
	(b) Sands	3 to 15	SW, SP, SM, SW-SM, SP-SM
F3	(a) Gravelly soils	> 20	GM, GC
	(b) Sands, except very fine silty sands	> 15	SM, SC
	(c) Clays, PI > 12	-	CL, CH
F4	(a) All silts	-	ML, MH
	(b) Very fine silty sands	> 15	SM
	(c) Clays, PI < 12	-	CL, CL-MI,
	(d) Varved clays and other fine-grained, banded sediments	-	CL and ML; CL, ML, and SM; CL, CH, and ML; CL, CH, ML, and SM

have higher bearing capacity during thaw than the F2 materials, even though they have equal volumes of segregated ice. Groups F3 and F4 show greatest weakening during thaw, while F4 soils are highly frost-susceptible.

In actual practice, some gravelly soils with about 1% material finer than 0.02 mm show significant heave, while some sandy materials having up to 20% of material finer than 0.02 mm are stable. Thus there is a need for laboratory testing of the "rate of heave" of soils in the laboratory under standard conditions in order to arrive at reasonably reliable conclusions on which specific critical designs can be based. A "classification of frost-susceptibility" based on laboratory heaving tests has been developed by the U.S. Corps of Engineers (Linell & Kaplar, 1959; Kaplar, 1974) to evaluate the results of these tests (Table 2.4). The tests are based on a laboratory frost penetration at a rate between 6 and 13 mm day^{-1}. Unfortunately, the tests take at least two weeks but Kaplar (1971) concluded that useful frost heave data can be obtained in two to three days if a rate of frost penetration between 76 and 200 mm day^{-1} is used.

Several factors actually affect the test results, of which freezing rate is probably the most critical (Penner, 1960). This is because the rate of heaving and the rate of frost penetration are inter-dependent. The use of the standardized frost penetration rate gives good comparative data for different materials but, for design purposes, the rate should be matched to that expected in the field for the best results. These rates of frost penetration encountered in the

Table 2.4: U.S. Corps of Engineers Classification of Frost Susceptibility, Based on Laboratory Heaving Tests.

Average rate of heave (mm/day)	Frost susceptibility classification
0.0 - 0.5	Negligible
0.5 - 1.0	Very low
1.0 - 2.0	Low
2.0 - 4.0	Medium
4.0 - 8.0	High
> 8.0	Very high

field vary considerably from place to place (G.H. Johnston et al., 1981, pp. 142-143).

Frost Heaving Pressures

Laboratory studies of frost heaving forces developed in widely differing materials suggest that particle size is a valid criterion of frost susceptibility (Penner, 1973). Everett (1961) predicted that ice lens growth in a saturated particulate material could be stopped by applying suction to the water phase, by applying pressure to the ice phase, or by a combination of both.

$$\Delta p_l + \Delta p_w = \frac{2\sigma_{iw}}{r_l}$$

where Δp_l = pressure on the ice ($N\ m^{-2}$)

Δp_w = suction pressure in the water ($N\ m^{-2}$)

σ_{iw} = ice-water interfacial energy ($J\ m^{-2}$)

r_l = pore radius (μm)

Since pore size distribution is related to grain size distribution and is an indicator of the permeability of the soil, it can be used to determine the heaving pressures that can be developed. Figure 2.22 shows the results obtained by Hoekstra et al. (1965) under conditions of minimal restraint. They indicate the variability between soils, especially in the fine sand - coarse silt range, due to differences in grain size distribution, as well as the high pressures produced.

Heaving pressures increase with increasing constraint (Kinosita, 1967; Penner, 1970; Yong & Osler, 1971; Radd & Oertle, 1973; Chen et al., 1983). However, in spite of all these tests and theories, the field situation shows a high degree of complexity and, as a result, variations in temperature, soil type, impurities in the pore water, groundwater locations, surcharge loading and surface cover make the prediction of actual frost heaving pressures too difficult to forecast accurately using existing methods.

Lambe and Kaplar (1971) found that additives can modify the frost susceptibility and heaving of soils. Unfortunately, most of these are expensive, are not readily available and may only provide a reduced heave rate.

Table 2.5: Terminology Describing Ground Temperature and Water Phases in a Soil/Water System (after van Everdingen, 1976, p. 864, fig. 2)

H_2O content → / Temperature ↓	No H_2O (except chemically bound and adsorbed)		Some H_2O (less than porosity)		Pore spaces filled with H_2O		Containing 'excess' H_2O		Zone descriptions: Phase	Zone descriptions: Temp.
T > 0°C	Dry, noncryotic		Moist or unsaturated, noncryotic		Wet or saturated, noncryotic		---		Nonfrozen	Noncryotic
0°C – Cryo point; T < 0°C	Dry, cryotic	Dry permafrost (1)	Moist or unsaturated, cryotic	Moist permafrost (1)	Wet or saturated, cryotic	Wet permafrost (1)	---	--	Nonfrozen	Cryotic
Initial freezing point of soil system; T < 0°C			Ice-poor, partially frozen	Ice-poor permafrost (1)	Partially frozen	Permafrost (1)	Ice-rich, partially frozen	Ice-rich permafrost (1)	Partially frozen	
			Ice-poor, frozen		Frozen		Ice-rich, frozen		Frozen	

(1) If temperature is perennially below 0°

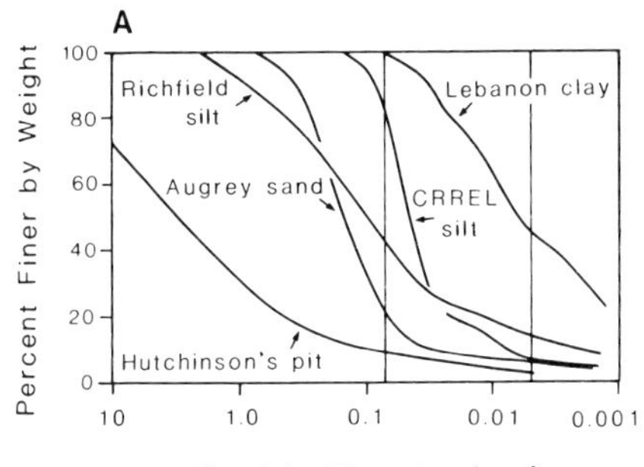

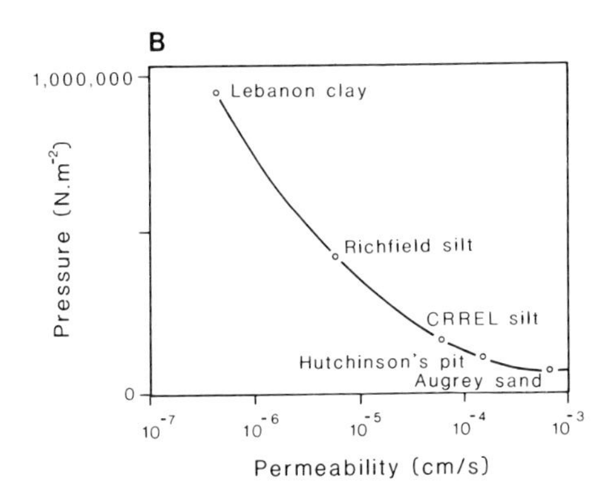

Figure 2.22: Heave Pressure Versus Permeability (After Hoekstra et al., 1965).

A specific case of the regular use of additives is reported by Voroshilov (1978), who has experimented with coagulators in north-east Siberia to minimize heaving. Although kaolinite gave 22.19 mm heave, this was reduced to 0.78 mm by a treatment of 1.5% by weight of sulphite alkalis and was reduced to 3.42 mm by a treatment of urea formaldehyde. However, the latter is a carcinogenic substance. This reduction in heave was found over a range of 18-41% moisture content. The sulphite alkalis were waste products from cellulose factories.

Cryotic Soils

Since all the water does not turn to ice at 0°C, and since some icy clays at -4°C contain so much liquid that they are plastic, a terminological problem becomes obvious. Van Everdingen (1976, 1985) has proposed additional terminology to cope with the problem (see Table 2.5 and Fig. 2.10B). Soil that is colder than 0°C is said to be "cryotic". This terminology will be followed in this book.

ADDITIONAL PROCESSES ASSOCIATED WITH FREEZING AND THAWING

There are several other important processes which take place in any material left in the zone of seasonal freezing and thawing that also act on man-made structures or foundations. These include seasonal frost heaving, upheaving of objects, upturning of objects, needle ice, downslope movements, sorting, contraction cracking, and frost comminution.

Seasonal Frost Heaving

This is the seasonal occurrence of frost heaving at a given location and Table 2.6 gives typical results for the surfaces of a variety of alpine and polar sites. There is little difference between high arctic sites on permafrost and low latitude sites in seasonal frost. Frost boils show the largest heaving values, while highly frost-susceptible silts and clays with abundant available moisture also show large amounts of heave. Heaving is least in dry, alpine sites, where it takes place after melting of the snow by chinook winds, resulting in sudden large inputs of moisture into the ground.

Heaving varies with depth (Chambers, 1967; Czeppe, 1959, 1960, 1966; Mackay et al., 1979), so that considerable stresses can be applied to any elongate object embedded in the ground. There is considerable variability in heave between centres and margins of sorted polygons (Chambers, 1967) and of earth hummocks (Mackay et al., 1979) and negative heave may occur at some locations in the microrelief. This indicates the generation of compensatory movements and pressures which Washburn (1956) has called cryostatic pressures.

Upfreezing of Objects

The upward movement of stones and other objects is often associated with ice segregation, both in the seasonal frost and in the active layer. This "upfreezing" can occur in sediments of all textures and can occur at the base or the top of the active layer (Corte, 1962a, 1963a; Mackay, 1973b, 1984; Mackay & Burrous, 1979; Washburn, 1947). It is responsible for the northern boundary of winter wheat and for the ejection of stones from farmers' fields. Tile drains can be crushed and ejected, and the pressures involved are sufficient to cause problems with culverts, piles, and brick, tile and concrete foundations. In the Arctic, artifacts of different ages are ejected to lie on the ground surface (D.L. Johnson & Hanson, 1974; D.L. Johnson et al., 1977), making stratigraphic study difficult.

The heaving is regarded as being predominantly autumnal in Greenland (Washburn, 1979, p. 86), and the upheaving during a single cycle is often proportional to the "effective height" of the object (Hamberg, 1915, p. 609), i.e. the vertical height of the stone above its greatest horizontal diameter.

Two mechanisms have been proven to operate in specific studies and they probably operate together in many cases. First is the frost-pull theory of Högbom (1910, pp. 53-54), whereby the slowly descending freezing-front causes ice lenses to develop in the soil around the top of the stone. Since the soil freezes to the top of the stone, the subsequent heaving of the soil lifts the stone upwards, leaving a space beneath it. During thawing, soil slumps into the space beneath the stone. Time-lapse photography confirms its effectiveness (Kaplar, 1965), while Kaplar (1969, p. 36) gives a formula for predicting the maximum distance of heave in one cycle.

The second mechanism is called the frost-push hypothesis (Högbom, 1914, p. 305). The thermal conductivity of stones is greater than that of soils, which causes ice to grow at the base of stones, pushing them upwards, when the surrounding fines are still unfrozen. Bowley and Burghardt (1971), Corte (1962d, 1963a), and Mackay and Burrous (1979) have demonstrated the process with downward and upward freezing in the laboratory.

Both processes may occur together and uplift of 5 cm would be a typical total movement for a year (Price, 1970). However, it needs to be borne in mind that wetting and drying can also cause upward movements of stones (Springer, 1958; Jessup, 1960).

Upturning of Objects

A characteristic of some permafrost environments is the upturning of many of the stones present so as to leave their long axes vertical. Vorndrang (1972) working in the Silvretta Alps, showed that in 46 freeze-thaw cycles (one season), 38% of the stones in a random mixture of soil had become vertically oriented, compared with 41% before mixing. Harris (1969) found that upturning of stones in till occurred in such a way that the intermediate axes of

Table 2.6: Frost-heave Values Obtained in Field Studies

Location	Climate	Period of Measurement	Site Characteristics	Total Heave Per Year (cm)	Reference
Cape Thompson, Alaska, 70°N	Polar Permafrost		Frost boil, very frost-susceptible	32.5	Everett, 1966
Signy Is., South Orkney Is., 61°S	Lowland Temperate-maritime Permafrost	1964	Sorted circle, highly frost-susceptible.		
			Surface	4.0	Chambers, 1967
			Buried	0.4	
			On stones	3.6	
			At edge	2.0	
Mesters Vig, Greenland 72°N	High Arctic Permafrost	1958-1964	Very frost-susceptible.		
			Wet slopes - 10 cm depth	0.0-1.0	Washburn, 1969
			- 20 cm depth	1.5-5.8	
			Dry slopes - 10 cm depth	0.5-0.7	
			- 20 cm depth	0.8-1.4	
Kananaskis, Alberta, 51°N	Dry Alpine Seasonal Frost	1970-1971	Frost-susceptible.		
			Wet slope, Dec. - Mch.	0.0-1.8	Harris, 1971
			Dry slope, Dec. - Mch.	0.0-0.4	
Colorado Front Range 39°N	Alpine Seasonal Frost	1969-1970	Highly frost-susceptible frost boil	25.0-29.5	Fahey, 1974
Colorado Front Range 39°N	Alpine Seasonal Frost	1965-1966	Very frost-susceptible. Stone-banked and turf-banked lobes	0.5-38.0	Benedict, 1970
Inuvik, N.W.T. 68°N	Lowland Permafrost under Forest	1976-1978	Highly frost-susceptible with clayey mud hummocks.		Mackay et al., 1979
			Undisturbed site (1977-8) Mean of 38 observations	10.39	
			Bulldozed site (1977-8) Mean of 9 observations	14.09	

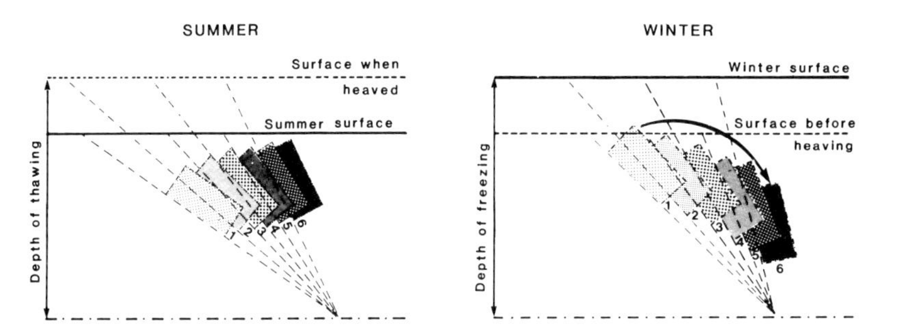

Figure 2.23: Diagram Showing the Process of Upturning of Stones

the pebbles ended up facing in the same direction as the long axes of pebbles in the underlying deposit, i.e., the mechanism is not random.

The mechanism most commonly suggested is a modification of the frost-pull theory (Fig. 2.23) whereby the inclined stone is lifted and rotated due to the resistance of the soil around the lower part of the stone (Pissart, 1969, 1973; Schmid, 1955; Washburn, 1979). In the alpine permafrost of Plateau Mountain, there was an average of 16 freeze-thaw cycles at 20 cm and 5 at 50 cm between 1976 and 1983, whereas in the high Arctic, only the annual cycle will be encountered below the surface 5 cm (Cook, 1966, p. 129). Presumably the process is far slower at high latitudes.

Needle Ice

Needle ice is the term used by Taber (1918, p. 262) for elongate ice crystals that grow beneath stones lying on the surface of the ground in maritime climates. It is discussed in detail in Washburn (1979). Lengths of 0.5 - 3.0 cm are normal, although needles up to 40 cm long have been reported (Krumme, 1935). The size depends on temperature, moisture and soil conditions. They grow on volcanic ejectamenta (Mount Garibaldi, British Columbia, and Iceland) or on stony loams, usually in a single night. Thawing next day causes the stone to slide off the ice pillar and move downslope. Cobbles weighing as much as 15 kg can be lifted (Mackay & Mathews, 1974b). This can produce significant sorting of material through differential heaving (Fahey, 1973; Gradwell, 1957; Hay, 1936; Troll, 1944, 1958), as well as downslope movement.

Downslope Movement

This can be found in natural landforms or in man-made deposits, especially under stress. Downslope processes in natural landforms can be divided into three components (frost creep, congelifluction and needle-ice creep) whose effects cannot readily be separated. Frost creep is defined as the net downslope displacement that occurs when a soil, during the freeze-thaw cycle, expands normal to the ground surface and settles in a nearly vertical direction. The best demonstration of its effectiveness with one-sided freezing is by Benedict (1970) and the process is discussed in detail in Washburn (1979). Mackay (1981) describes two-sided heaving during summer due to moisture leaving the thawed active layer and entering the permafrost and causing additional plug-like downslope movement of earth hummocks on Garry Island.

The term 'congelifluction' was introduced by Dylik (1951) for soil flowage over permafrost. Many workers, e.g. French (1976),prefer to include it with gelifluction, i.e., the downslope soil flowage over seasonal frost, but there is a case to be made for retaining the term. Gelifluction occurs in the spring when melted snow joins thawed surface soil in producing a layer with high pore-water pressure, which then flows slowly a short distance downslope. The soil then dries and movement ceases. Mackay (1981) has described evidence for summer flowage of thawed peat on 1-3° slopes, resulting from water additions by snow-melt and summer rains, together with strong through-flow from higher ground. Wu (1984) has shown that the tilted trees on 30° slopes on Poker Creek near Fairbanks are also due to congelifluction. Roots tend to be confined to the moss and Ah horizons, and show that the movements are compensated for by elongation of the roots downslope. Up to 8 cm of movement occurred in the moss layer in a single summer, with a maximum concentrated in the early summer thaw. From the nature of the perched water observed in the active layers over permafrost elsewhere (Fig. 2.16), this process is expected to prove to be rather widespread.

Needle-ice creep is easier to differentiate in that it is a downslope movement resulting from the growth and subsequent melting of needle ice in a maritime climate. It has been described by Ellenberg (1974) and Mackay and Mathews (1974a, 1974b) and may result in differential lobal movement of fine and coarse material. It only affects the upper 7 cm of material, but it can move in lobes of up to 7 m in four years (Mackay & Mathews, 1974b).

In addition, natural and man-made frozen materials can permit grains to move in them (Gold, 1985) and are also subject to long-term creep in the frozen state (Ladanyi, 1972). Morgenstern (1985) summarizes the present knowledge on the prediction of this phenomenon. Stresses can increase the rate and degree of flow, and a given soil has greater strength at lower temperatures. This will be discussed further in Chapter 4.

Sorting

This is discussed *in extenso* by Washburn (1956, 1973, 1979). It can be caused by needle ice and by upfreezing of stones as already noted. However, it can also be produced by mass displacements of material and results in the formation of sorted patterns on the ground surface. Lötschert (1972, p. 7) described plants from the margins of sorted circles with tap roots which could be traced horizontally to the centre of the circle. Schunke (1975a, p. 63) has described stones sliding downslope to the stony borders during thawing of the ground. Sorting by slopewash flowing in the stony channels has been described by Czeppe (1961). Nansen (1922) has suggested that differential weathering can simulate sorting. Miller (1984) discusses the evidence for thermally induced regelation which can permit movement of grains through ice.

In numerous studies, Corte (1961a, 1962a, 1962b, 1962c, 1962d, 1962d, 1962e, 1963a, 1963b, 1966a, 1966b) and others have concluded that under the right conditions, finer material migrates ahead of the freezing front, while the coarser material is more readily anchored by the freezing plane. Factors include orientation of the freezing front, rate of freezing, moisture content, shape, size, and possibly density of the mineral grains. It can produce both horizontal and vertical sorting. Ice can also transport particles upwards under a freezing gradient towards the warm side (Hoekstra & Miller, 1965; Radd & Oertle, 1973; Römkens, 1969; Römkens &

Miller, 1973), while clay could move with water to the freezing front (Thoroddsen, 1913, 1914; Johansson, 1914; Cook, 1956; Washburn, 1956; Schunke, 1975; Brewer & Haldane, 1957).

Whatever the cause, sorting (Plate IV) can be very rapid under the right circumstances (Strömquist, 1973; Konishev et al., 1973), and can operate on man-made structures such as gravel pads and gravel islands.

Contraction Cracking

One of the features of polar regions are the loud booming noises heard periodically during winter. They are caused by cracking of the ground in response to cooling and contracting after the ground has become brittle (Leffingwell, 1915; Lachenbruch, 1962). They require a minimum of 1000 °C days freezing index (Harris, 1982a), but are not true zonal permafrost features. Instead, they can occur in exceptional winters where permafrost is absent. They form at sites with higher thawing indices in peat than in mineral soil, and form the basic cracks that permit the growth of sand and ice wedges. In fall, the contraction due to cooling of the ground is probably partly offset by expansion of water as it forms ice. Once the ground becomes solid and brittle, fracturing can occur.

Studies by Mackay (1974) and Mackay and Mackay (1974) suggest that the bulk of the cracking in the ice wedge polygons of the western Arctic occurs between mid-January and mid-March. At Garry Island, at a MAST of -7°C, nearly 40% of the ice wedges cracked, whereas at Barrow (MAST = -9°C), about 50% crack (Black, 1963, 1974). Crack frequency depends on depth of snow cover (Mackay, 1978), one metre of snow being the critical snow depth to prevent cracking on Garry Island. Cracking commences in the ground and spreads upwards and downwards to a total vertical extent of up to 5 m. Crack width at the surface averages 1 cm. Where ice and sand wedges are absent, cracking does not necessarily occur in the same place every year.

Cracking occurs on artificial islands, gravel roads and building pads, and can break any buried cables it intersects. Accordingly, most telephone and power cables are located above ground in permafrost areas, except in special situations, e.g., alongside runways.

Frost Comminution

This is a convenient term to include frost wedging and hydration shattering. Frost wedging (Middleton, 1743, pp. 158-9) is the breaking apart of materials, commonly rock, by expansion of water during freezing. The most obvious source is the 9% expansion of water when changing to ice, but it can also be due to directional growth of ice crystals in the rock. Certainly ice can generate pressures up to nine times the average maximum tensile strength of unfractured rock (cf. Bridgeman, 1912, pp. 473-484; Grawe, 1936; Tricart, 1970) but it appears to be a very complicated process when studied in detail (see the detailed discussion in Washburn, 1979).

Hydration shattering is the ordering and disordering of water molecules in sorptive interactions with water vapour and liquid water, and it can produce shattering in shales, siltstones and argillaceous carbonate rocks (Dunn & Hudec, 1965, 1966, 1972; Hudec, 1974). Sorption and desorption of unfrozen water cause expansion and contraction, resulting in fatigue in these sorption-sensitive rocks. In all probability, true frost wedging only occurs in coarse-grained rocks, while most frost comminution is due to a combination of frost wedging and hydration shattering. The deeper layers of thick block field deposits such as the 40 m thick rubble on Plateau Mountain must be produced primarily by hydration shattering since the active layer is never more than about 15 m deep (Harris & Brown, 1978).

Both processes produce material obeying Rosin's Law of Crushing, but experiments by Konishchev (1973, 1978) and Konischchev et al. (1975) suggest that hydration shattering produces smaller particles, unlike the coarse blocks resulting from frost wedging, although Potts (1970) and Guillien and Lautridou (1974) give examples where silt and clay are not the end-products of hydration shattering. White (1976a, 1976b) has suggested that hydration shattering operates on block slopes and in most mass wasting phenomena. Subsequent chemical weathering of the fractured rock tends to produce clay-sized minerals (B. Meyer in Semmel, 1969; Leshehikov & Ryashchenko, 1973, 1978).

Frost comminution alters the size of material in pads, road beds, etc., and in extreme cases, creates considerable problems. Thus, the siltstone common on the east side of the Richardson Mountains in the Northwest Territories quickly disintegrates into

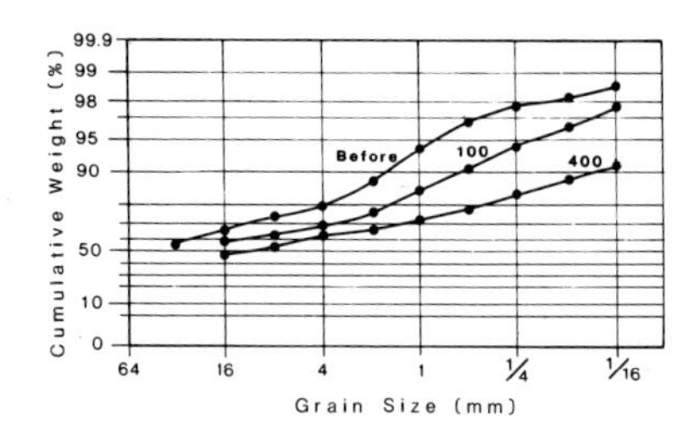

Figure 2.24: Disintegration of a Rock Sample from Willow Fan Gulley near Aklavik, Northwest Territories, by Freeze-Thaw Cycling in the Laboratory (data from Legget et al., 1966, plotted on Rosin Law paper).

silt when exposed to frost comminution (Fig. 2.24; data from Legget et al., 1966). Since this is the main rock available for borrow in the area, this section of the Dempster Highway is extremely slippery when wet.

Some idea of the relative effectiveness of frost comminution in different climates may be found in rockfall production (Table 2.7). Highest rates are found in maritime, alpine climates, while maritime arctic and continental alpine climates appear to produce an order of magnitude less debris. Continental arctic, alpine climates may produce even three orders of magnitude less material than maritime, alpine climates. Thus, the climate is clearly a very significant factor in rates of frost comminution, although lithology is also very important.

Table 2.7: Rates of Production of Rockfall Debris in Arctic and Alpine Environments from Steep Cliffs

Location	Climate	Rock type	Rate (mm yr^{-1})	Author
Swiss Alps	Maritime, alpine		2.5	Barsch, 1977
Longyeardalen, Spitsbergen	Maritime, arctic		0.3	Jahn, 1976
Surprise Valley, Jasper National Park	Alpine, continental	Limestone	0.06 - 0.26	Luckman, 1972
Ogilvie & Wernecke Mtns., Yukon	Continental, arctic, alpine	Syenite and Monzonite	0.003 - 0.019	Gray, 1973

Chapter Three

DISTRIBUTION AND STABILITY OF PERMAFROST

Figure 1.1 demonstrates that permafrost is widespread in the polar and high altitude regions on the major continents in the northern hemisphere where the mean annual air temperatures are sufficiently cold. Permafrost also occurs on the shallower polar sea floors, e.g. the Beaufort and Laptev Seas. In this chapter, the zonation, distribution and relative stability of the areas underlain by permaforst will be discussed. Also included is a summary of the main controlling factors producing these effects. Knowledge of these factors is fundamental to the successful use of permafrost regions by man if environmental impacts are to be minimized.

ZONATION OF PERMAFROST

It is generally agreed that permafrost occurs in two major situations, viz.: polar permafrost and alpine permafrost. Polar permafrost has been studied most extensively, but even there, no simple consensus exists on how to subdivide it (see for example, Heginbottom, 1984; Kudryavtsev et al., 1978, 1980; Kreig, 1985; and Washburn, 1979). This is due to differences in the scale of the maps and the types, qualities and quantities of information available (Harris, 1986).

The most detailed quantitative classifications are used in the USSR, where up to six subdivisions may be used, based primarily on the percentage of the earth's surface (including lakes and sea floor) that is underlain by permafrost (State Committee . . . , 1960, in Brown, 1963, p. viii; Kudryavtsev et al., 1978, 1980). This greater detail is made possible by the intensive permafrost studies in areas undergoing development. However, in North America and China, this detailed classification is not

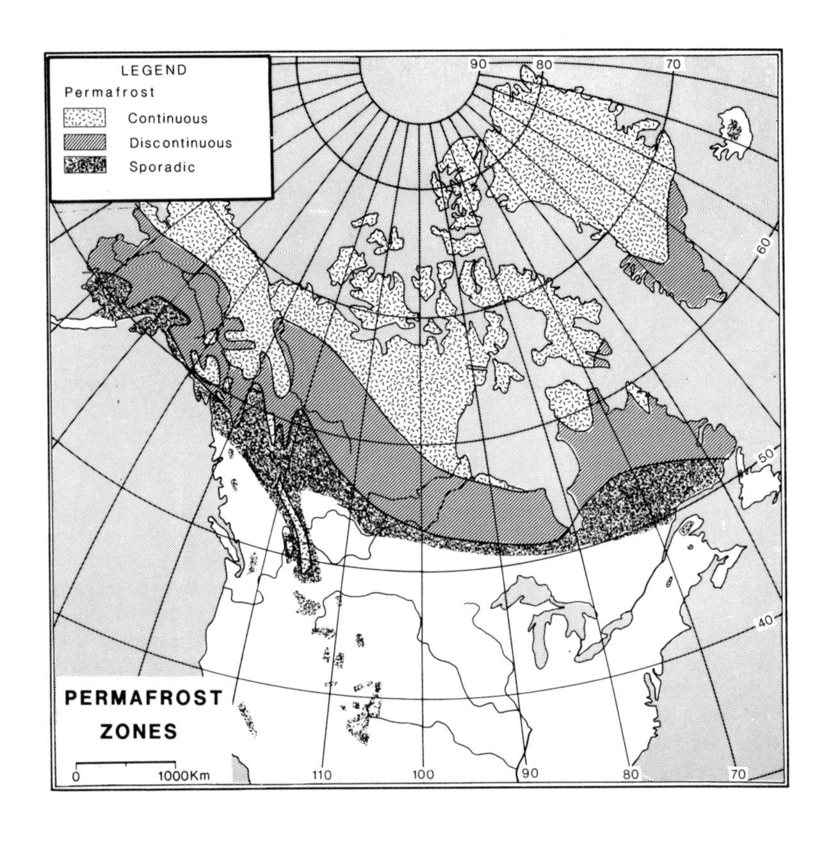

Figure 3.1: Distribution of Permafrost in North America (modified from Harris, 1985).

feasible due to limited availability of information.

In this book, the broad subdivision by the Russians (Kudryavtsev et al., 1978) into continuous permafrost (>80% underlain by permafrost) and sporadic permafrost (<30% underlain by permafrost) will be used, since it forms a good basis for world-wide maps even in areas with limited factual data. The term 'discontinuous permafrost' will be used for the intermediate category (30-80% underlain by permafrost). A more detailed discussion is given in Harris (1986).

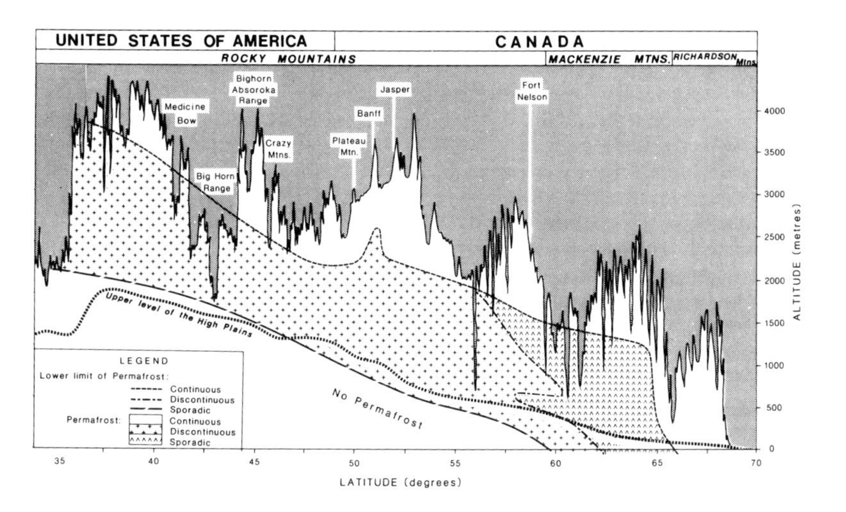

Figure 3.2: Permafrost Zonation Along the Eastern Ranges of the Rocky Mountains.

DISTRIBUTION OF PERMAFROST

It will be seen from Figure 1.1 that the largest areas of permafrost in the northern hemisphere are located in eastern Asia and in Canada. Permafrost extends southwards along the higher mountain chains in the Western Cordillera of North America and in the highlands of Mongolia and Tibet.

Figure 3.1 shows the distribution in North America, based on Brown (1967, 1968), Ferrians (1965a, 1965b, 1966), Ferrians and Hobson (1973) and Harris (1983, 1985). The thickness of the continuous polar permafrost varies from 60-90 m at the southern limit to over 500 m in the Arctic Islands (Judge, 1973a). It may exceed 1000 m thickness in the mountains of Ellesmere and Baffin Islands (Brown, 1972). Lowest mean annual air temperatures (-20°C) and mean ground surface temperatures (-17°C) occur at the north end of Ellesmere Island, compared with about -5°C at the boundary with discontinuous permafrost. Permafrost is about 650 m thick at Prudhoe Bay (Lachenbruch & Marshall, 1969; Gold & Lachenbruch, 1973). The considerable variability in thickness elsewhere along the north coast (Fig. 2.8) has been attributed to variations in thermal conductivity of the bedrock (Lachenbruch, 1970). Southwards, there is a transition through the broad discontinuous permafrost to the zone of

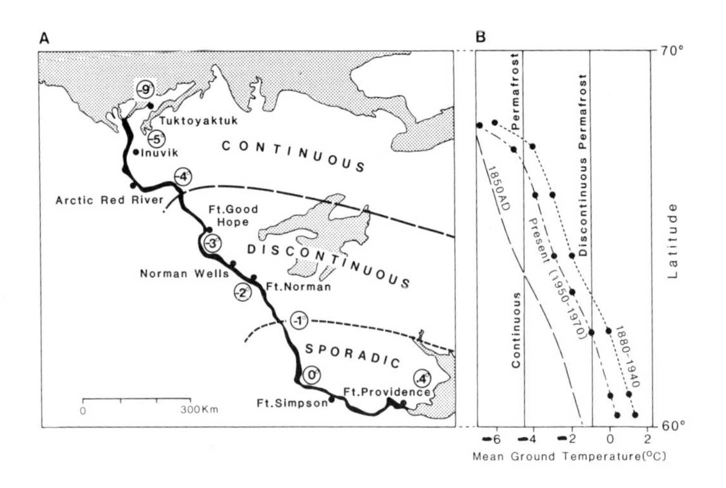

Figure 3.3: A. Approximate Present-day Mean Annual Ground Temperatures in °C (after Mackay, 1975) Along the Mackenzie Valley. B. Changes in Ground Temperature Inferred from Changes in the Active Layer Thickness, Ice Wedges (Mackay, 1975, 1976) and from Recent Climatic Data (Burns, 1973).

sporadic permafrost. Taliks become larger, the active layer thickens from less than 1 m to at least 3 m and the permafrost thins to about 12 m (Brown, 1970a). In the sporadic zone, permafrost is represented by ice beneath peat and by buried, isolated ice masses. The ground temperatures in these patches of permafrost are generally above -1°C.

Along the eastern ranges of the Western Cordillera (Harris, 1985, 1986), the same zones can be traced southwards to about 35°N (Fig. 3.2). However, the zone of discontinuous permafrost becomes reduced to about 70 m in altitudinal extent while the zone of sporadic permafrost increases to a vertical range of 1000 m, probably due to better drainage in the alpine zone.

Since most of the North American permafrost areas were covered by ice during the glacial periods, most of the permafrost has been developed during the Holocene period. Even so, fluctuating climate has caused appreciable areas of disequilibrium permafrost (see Chapter 2). Thus, along the Mackenzie valley (Fig. 3.3), small changes in the winter air temperature have resulted in substantial north-south changes in the distribution of permafrost and ground ice in the last 150 years (Mackay, 1975). Since 1940, the temperature changes have caused a 10-40% decrease in the thickness of the active layer in the Mackenzie delta, resulting in the development of secondary and tertiary ice wedges growing on top of the old wedges (Mackay, 1976). Changes along the southern boundary of the zones have resulted in the local development of thermokarst and in the disappearance of ice wedges, palsas and peat plateaus from Alaska (Péwé, 1966b), across the northern prairie provinces (Zoltai, 1971) to Quebec (Ives, 1979; Lagerac, 1982).

The eastern part of the USSR is largely underlain by permafrost (Fig. 3.4), which continues southwards into Mongolia, northern China and Tibet. The maps of Kudryavtsev et al. (1978, 1980), Baranov (1959, 1964) and Shi Yafeng and Mi Dasheng (1983) provide a good source for additional detail.

Much of eastern Asia was not glaciated during the Pleistocene, so that much of the permafrost is relict (Baulin, 1962; Fotiyev et al., 1974; Oberman, 1974) and exceptionally thick beneath the old land surfaces. Maximum thicknesses reported are 1400-1450 m in the vicinity of the upper Markha river (Grave, 1968a, 1968b), although the lowest mean ground surface temperatures are only about -12°C. The relict permafrost is often very rich in ice and has exceptionally large ice wedges up to 30 m deep. These are believed to have formed during the glacial period and Balobayev (1978) has calculated that the air temperatures were 10°-13°C lower than today about 20,000 years ago, based on permafrost thicknesses and heat-flow theory. As in the case of North America, recent cooling of the climate has resulted in thinning of the active layer during the last few decades (Belopukhova, 1973). Pleistocene permafrost is also found under the shallow seas along the Arctic Ocean, e.g. beneath the Laptev Sea.

The relative aridity of the permafrost region and the deep freezing have produced pockets of cryotic liquid saline water (cryopegs) in the ground below large areas. These brines can move and penetrate other layers.

In the southern alpine permafrost region, the history is more complex. During the early Pleistocene, the land was not as high as at present and warm Monsoon winds brought moisture from the Indian Ocean, producing extensive glaciers on the mountains of Tibet (Gerasimov and Zimina, 1968; Shi Yafeng et al., 1979). Presumably permafrost did not extend as far south as today. When the Himalayas

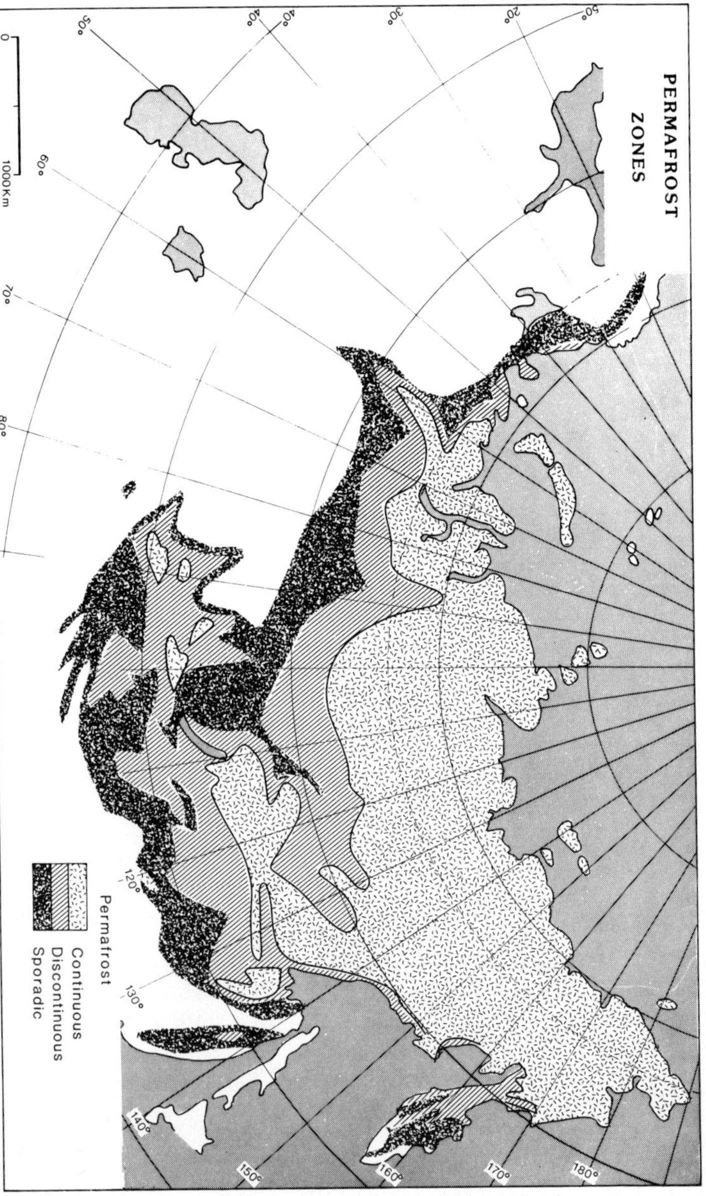

Figure 3.4: Distribution of Permafrost in Asia (after Kudryavtsev et al., 1978, 1980; Brown, J., & Yin-Chao Yan, 1982).

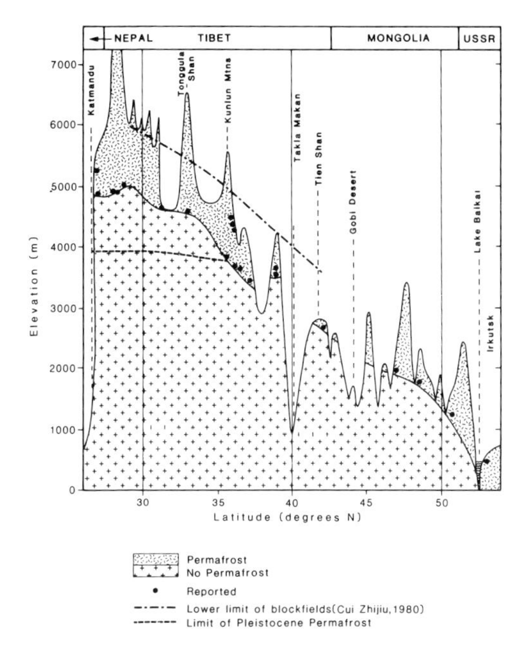

Figure 3.5: Lower Limit of Permafrost with Latitude Across the Tibetan Plateau and Himalayas from South-Central Siberia (data from Fujii & Higuchi, 1976; Hsieh et al., 1975; Fujii & Higuchi, 1978; Froehlich & Słupik, 1978; R.I.G.C.D.R., 1975; Chou & Tu, 1963; Cheng Guodong, 1984; Cui Zhijiu, 1980).

cut off this moisture source in the middle Pleistocene, the glaciers disappeared and permafrost became established across the region as the Siberian cold winter air mass became dominant. Today, deep relict permafrost is found in these older glacial deposits (Fig. 3.5). Cold air drainage downslope results in continuous permafrost north of 50°N, but it is thinner than in north-east Siberia due to the shorter time for its development (see Baranov, 1959, 1964).

The permafrost areas of western China, southern Mongolia and Tibet are unusual in that they combine aridity and alpine permafrost. Active pingos occur along drainage lines and around playa lakes, together with saline and alkali Brown Desert soils

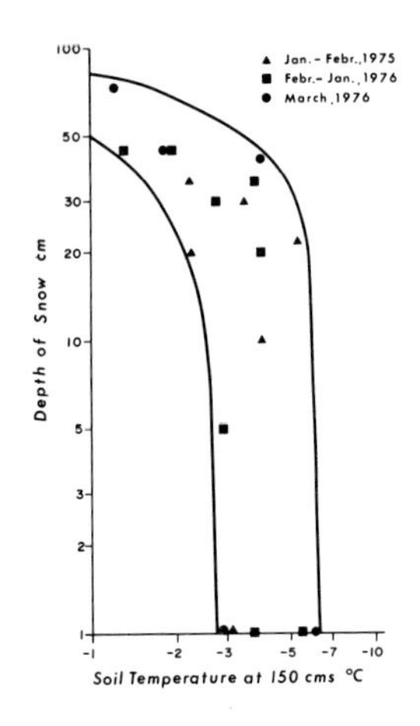

Figure 3.6: Effect of Mean Snow Depth in Winter on the Soil Temperature at 150 cm on Plateau Mountain, Alberta (from Harris & Brown, 1978). Reproduced with the permission of the National Research Council of Canada.

(Kowalkowski, 1978). Rock glaciers occur on the mountain slopes (Cui Zhijiu, 1983).

Arid polar permafrost occurs in the eastern Arctic Islands of Canada and in parts of Antarctica. The other extreme is the maritime polar permafrost of Spitsbergen (Åkerman, 1980; Corbel, 1956; Dutkiewicz, 1967) and the maritime alpine permafrost of Iceland (Schunke, 1975). These are characterized by patterned ground and gelifluction landforms.

FACTORS AFFECTING PERMAFROST DISTRIBUTION

Brown and Péwé (1973) divided these into climatic and terrain factors, although they overlap. The climatic factors control the level (temperature) and amount (duration) of heat applied to the ground surface. Altitude, latitude and snow cover are involved. The effect of the climatic factors is modified by the terrain factors, e.g. local relief, vegetation, hydrology, nature of the soil or rock, and fire. In addition, time and the activities of man are involved.

Generally, terrain factors, together with time and man, control permafrost distribution and thickness in the zones of discontinuous and sporadic permafrost, whereas the climatic factors are dominant in the areas of continuous permafrost (Brown, 1970a, p. 22).

Temperature

As discussed in Chapters 1 and 2, permafrost is a thermal condition (see Gold & Lachenbruch, 1973). Although air temperature provides a measure of heat received from advection and radiation, it does not have a consistent relationship with ground temperature (Fig. 2.4) due to the interference of the other factors (Brown, 1960). Not the least of the problem is that climate is not constant but is continuously changing to some degree.

Although mean annual air temperature (MAAT) does not correlate too well with permafrost distribution, greater success is obtained with total annual freezing and thawing indices (see Fig. 2.5), as discussed in Chapter 2.

Snowfall

The effect of winter snowcover as an insulator, protecting the underlying ground from the cold air above, is well established (Brown, 1966b; Mackay, 1978; Nicholson, 1978a). Figure 3.6 shows the effect at Plateau Mountain, Alberta (Harris & Brown, 1978) where 50 cm is the critical lower thickness to prevent freezing of the ground (total annual freezing index about 2000 degree days C/year).

However, this critical depth will vary with total annual freezing index and with the insulating properties of the snow cover. Thus Nicholson and Granberg (1973) found 65-75 cm to be the critical snow depth at Schefferville, Quebec, while it is >100 cm in the Arctic Islands of Canada (A.S.Judge, personal communication, 1981). On a continental scale, the differences in the distribution of permafrost on the two sides of Hudson Bay (Fig. 3.1) are due to the deeper snow cover in Quebec.

Duration of snow cover is also important, since the persistence of snow reduces the time the ground is subject to thawing and summer heating. Together with time of arrival of the snow cover, drifting and variations in snow density, this can cause tremendous variations in ground temperatures over a short distance (Brown, 1972; Seppälä, 1982). Summer snowfall greatly reduces the effect of summer heating (French, 1970). The first Fall (autumn) snowfall removes the remaining heat from the surface of the soil by using it to supply the latent heat of fusion of the ice crystals.

Altitude and Latitude

The general effect of latitude and altitude on permafrost is demonstrated in Figure 3.2, but

Figure 2.4 demonstrates that the terrain factors cause considerable variation in temperature in a given situation. Latitude operates mainly through the change in angle of incidence of the solar radiation and by controlling the duration of insolation. In mid-summer, the angle of the sun's rays is much lower in polar areas, so that the actual potential radiation received on a horizontal surface in 24 hours is reasonably constant from 52° to 90°N latitude. In mid-winter, there is no sunshine north of the Arctic Circle, whereas there is still 8 hours' daylight at 52°N. When freezing and thawing indices are plotted against altitude or latitude, a good relationship is demonstrated, although it is usually modified abruptly by other factors such as vegetation cover (Fig. 3.7).

Cold Air Drainage

Very large modifications of the applied cold can occur in mountainous terrain due to drainage of cold air downslope under otherwise stable anticyclonic conditions with clear skies and intense heat loss by reradiation (Harris, 1983b). These

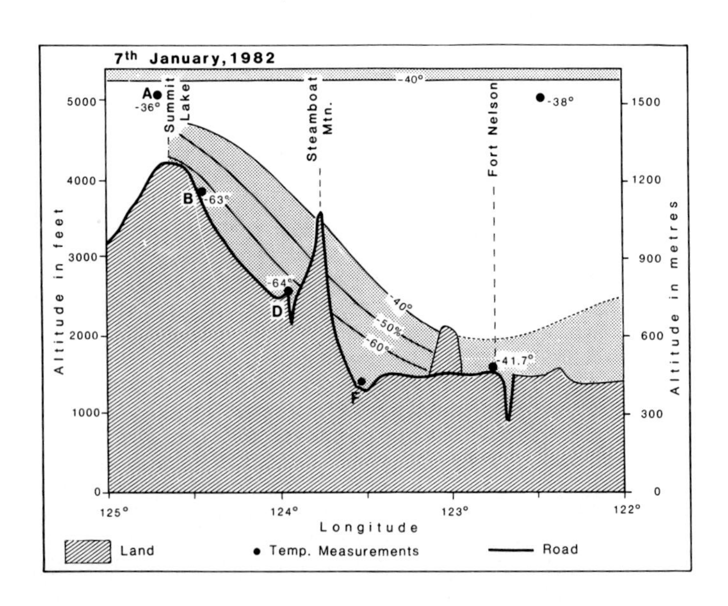

Figure 3.8: Temperatures Along the Alaska Highway During a Cold Air Drainage Event (modified from Harris, 1983b).

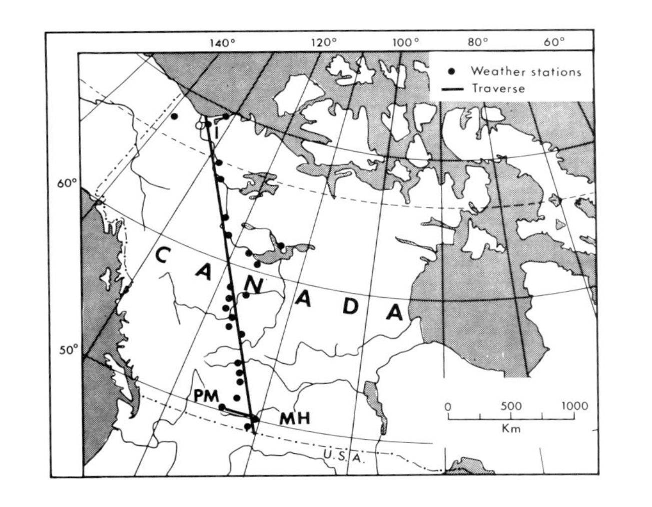

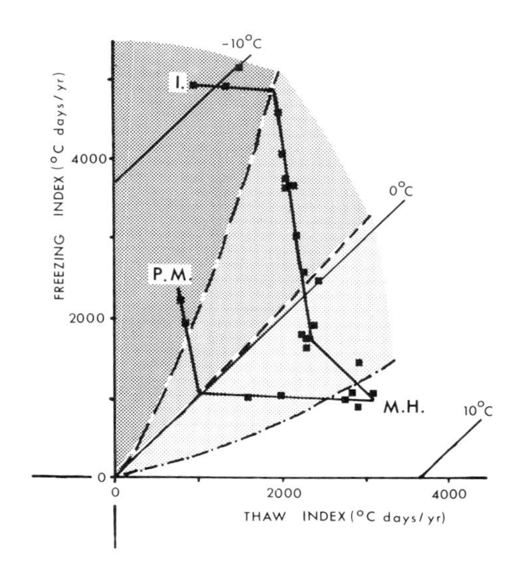

Figure 3.7: Relationship Between Freezing and Thawing Indices, Latitude and Altitude Along Two Traverses from Medicine Hat in Western Canada (from Harris, 1981a, 1981b).

provide the extreme low winter temperatures recorded at the Department of Transport Class A weather stations at Snag and Mayo, Yukon Territory, and the lowest temperature measured so far in North America (-64°C after correction) in a valley in the foothills along the Alaska Highway west of Fort Nelson in January, 1982 (Fig. 3.8).

Topography

This determines the microclimate by modifying the amount of solar radiation received by the ground surface in hilly situations. Orientation and slope angle are the key factors and are especially important between the latitudes of 55° and 66° and in mountains. In polar situations, the 24 hours of insolation in summer partially overcomes this effect. In areas of continuous permafrost, the active layer is normally thinner and the permafrost thicker on poleward-facing slopes (Brown and Péwé, 1973), while in areas of discontinuous permafrost, the equatorial-facing slopes tend to be underlain by taliks (Brown, 1969; G. H. Johnston, 1981, Fig. 2.4). These relationships can be modified locally by other factors (see French, 1970, who reported thinner active layers on southwest slopes on the Beaufort Plain, northwest Banks Island, due to evaporative cooling).

At the microscale, topography causes local modifications in the vegetation, winter snow cover and hydrology, which, in turn, modify active layer thickness and distribution of taliks. These are a prime cause of small frost mounds.

Mountains often cause local sun traps and cold spots, as well as inversions and cold air drainage. There is usually an inversion located about 100 m above a col or saddle in a pass (Harris & Brown, 1982), while the extreme case of cold air drainage is probably that of the cold air descending from the Tibetan Plateau northwards downslope along the major rivers such as the Lena river past Irkutsk and Yakutsk.

Vegetation

The vegetation cover modifies the micro-climate, although this effect decreases in the area of continuous permafrost and is virtually absent in the polar deserts (Brown, 1963,1970, 1972). The most obvious effect is in reducing the depth of thaw of the active layer by shielding the ground from insolation (Lindsay and Odynsky, 1965; Brown and Johnson, 1965; Jeffrey, 1967; Price, 1971). Moss thickness and shading by trees are key factors (Greene, 1983).

Removal or disturbance of the surface cover usually causes degradation of the underlying permafrost (Mackay, 1970), unless moss and peat are left protecting the ground (Linell & Johnston, 1973). Peat is a special case and has been studied intensively (Brown, 1968, 1969). In summer, the peat dries out and becomes an excellent insulator. Autumn snows wet the peat which then freezes, producing a layer with relatively high thermal conductivity. This results in near-surface permafrost appearing in thick peats on the southern margins of permafrost.

Trees shade the ground from solar radiation in summer and act as radiators of heat in winter. Conifers intercept some of the snow, altering the heat exchange significantly (Viereck, 1965). They modify wind velocities at the ground surface, reducing heat transfer. In peatlands with stunted trees, air temperatures are several degrees colder than in other vegetation types (Williams & van Everdingen, 1973). Coniferous and deciduous trees have different effects on albedo and evapotranspiration, especially in winter, the coniferous forest having about 1°C higher air temperatures.

Seral changes in vegetation are accompanied by changes in the permafrost distribution. Thus colonization of vegetation along rivers is usually accompanied by increasing development of permafrost, e.g., on the Mackenzie Delta (Smith, 1975), but in interior Alaska, permafrost is more widespread in the early successional stages of willow and balsam poplar (Van Cleve & Viereck, 1983; Viereck, 1970; Péwé, 1970).

Hydrology

Moving water is an effective thawing and erosive agent, particularly in ice-rich ground (Mackay & Black, 1973). This is particularly well demonstrated after fire or other removal of vegetation.

Where lakes or rivers do not freeze to the bottom, a talik will be present. Large, shallow water bodies, high water temperature, thinner winter ice and thick snow cover favour a large talik (Feulner & Williams, 1967), as do dark-coloured bottom sediments (Brown & Péwé, 1973). Lakes and rivers can cause thawing of the surrounding sediments, which can result in enlargement of the water body in ice-rich sediments.

In arid regions such as Tibet, lines of water flow into inland drainage areas, creating pingos and icings (Kowalkowski, 1978; Froehlich & Słupik, 1978). In wetter areas, they tend to produce seasonal frost blisters, East Greenland-type pingos and icings. Underground drainage lines usually represent the last taliks in an area and the permafrost severely limits the locations at which water can come to the surface.

Nature of the Soil and Rock

There can be tremendous variation in the albedo and thermal properties of soil, rock or snow, which can produce considerable local variability in permafrost distribution. Dark colours absorb more incoming radiation. Thus reflectivity may be 90% for snow, 12-15% for rock and 15-40% for bare soil (Brown & Péwé, 1973; Brown, 1973a).

Moisture content affects the amount of heat used for evaporation, while the nature of the material determines its ability to store water. The moisture greatly alters the freezing rate and the nature of the freezing process.

Since geothermal heat flow is fairly constant, the thermal conductivity of the substrate determines the thickness of permafrost at a given site (see Fig. 2.8), and also the thickness of the active layer (Brown, 1973b; Tedrow, 1966; Day & Rice, 1964). On Devon Island, Peters and White (1971, in Brown & Péwé, 1973, p. 79) reported active layer thicknesses of 70 cm in Stony Arctic Brown soils on gravelly beach ridges, 47 cm in Regosols, 34 cm in Arctic Brown soils, 24 cm in Gleysols, and 13 cm in peats.

Fire

Many natural fires are started each year within the permafrost areas, especially in the forested zones (Viereck, 1973; Sykes, 1971). Most of the discontinuous permafrost zone has been burned over at least once during the Holocene. Fire can smoulder for months during summer in the spruce-lichen woodlands of the forest-tundra transition zone (Rouse & Kershaw, 1971), but in wet tundra, there is less woody fuel available and the effects are greatly reduced (Johnson & Viereck, 1983; Wein, 1976).

Fast-moving fires burn the trees over wet ground but only the surface vegetation is removed, allowing peat plateaus and palsas to survive. Thus, in the Hudson Bay Lowland and northwestern Manitoba, only the top 2.5 cm of the vegetation mat is affected by fires and the underlying peat protects the icy permafrost (Brown, 1965, 1968).

Where conditions are dry, e.g. at Inuvik on 11th August, 1968, the vegetation and litter is completely burned, leaving blackened soil at the surface. Ice-rich frozen ground then thaws, resulting in surface meltwater and considerable thermal erosion by freeze-up in mid September. Firebreaks bulldozed to the permafrost table to control the fire suffer even greater thermal erosion. Study in subsequent years shows that the increase in depth of thaw is greatest in the fire-breaks and continues deepening for at least eight years, even when a shrub vegetation has been regenerated (Heginbottom, 1970, 1971, 1973; Mackay, 1977). Similar effects have been reported from Alaska (Viereck, 1982).

Glaciers

Glacier ice acts as a blanket covering the ground underneath it but it differs from snow in having a much higher thermal conductivity. As a result, the ice can sustain a thin zone with negative temperatures beneath it (cold glaciers) if other conditions are suitable. Beneath warm glaciers (i.e., the base of the glacier is at pressure-freezing point), no permafrost will be found. Tests beneath the Greenland ice sheet, 150 km east of Thule showed a bottom temperature of -13°C beneath 1387 m of ice (Hansen & Langway, 1966; Langway, 1967; Robin, 1972) but warm glaciers may occur even in the high Arctic (Müller, 1963).

Unfortunately, there are few data from most glaciers and the available data suggest that the temperature regimes beneath an ice-cap are quite complex and varied (Philberth & Federer, 1971). Judge (1973a) suggests that the present ground surface temperatures in northern Canada are lower than the ice-bottom temperatures during the Pleistocene, based on an ice-cap thicker than 1200 m. During deglaciation, inundation by proglacial lakes would probably have thawed any permafrost that was present. However, areas not protected by Pleistocene ice sheets developed deep, cold permafrost which could not form under the present climate.

Time

Long periods of time are taken to achieve an equilibrium under any given set of conditions and this must be taken into account in predicting the effects of environmental interference by man or by nature, e.g. climatic change.

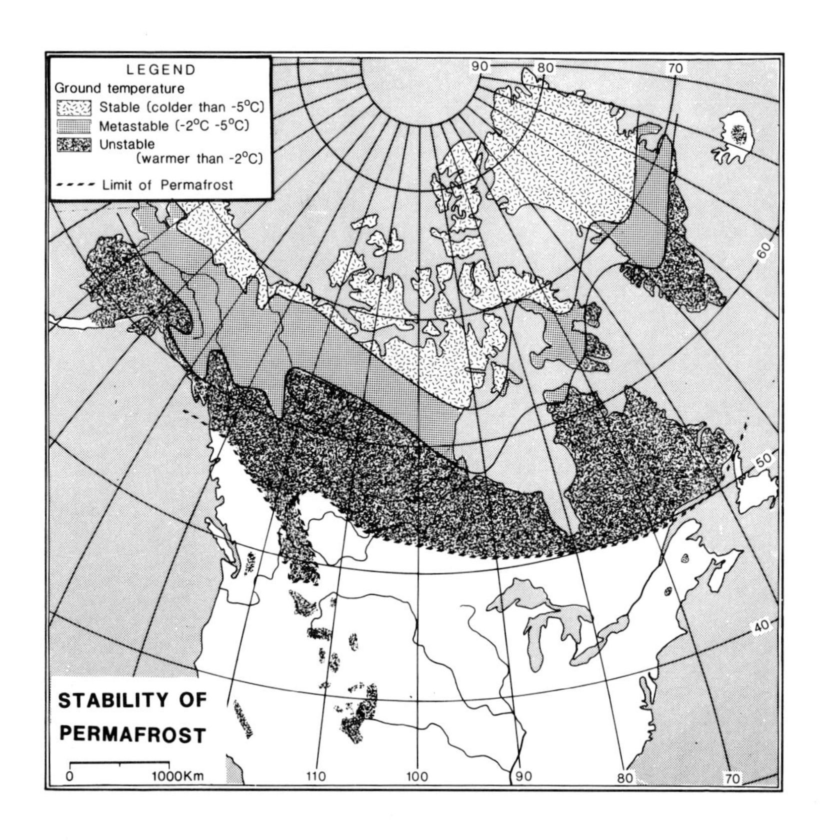

Figure 3.9: Thermal Stability of Permafrost in North America.

Man

Almost all of the works of man in permafrost areas that will be discussed in succeeding chapters result in changes in the thermal regime. As more use is made of permafrost areas, man may well become the major modifier of the climatic factors.

STABILITY OF THE PERMAFROST

This discussion of the factors affecting permafrost indicates that even without interference by man, there is a probability that any given permafrost area will, from time to time, be subjected to a change in its environmental conditions. It is therefore desirable to have a system of classifying

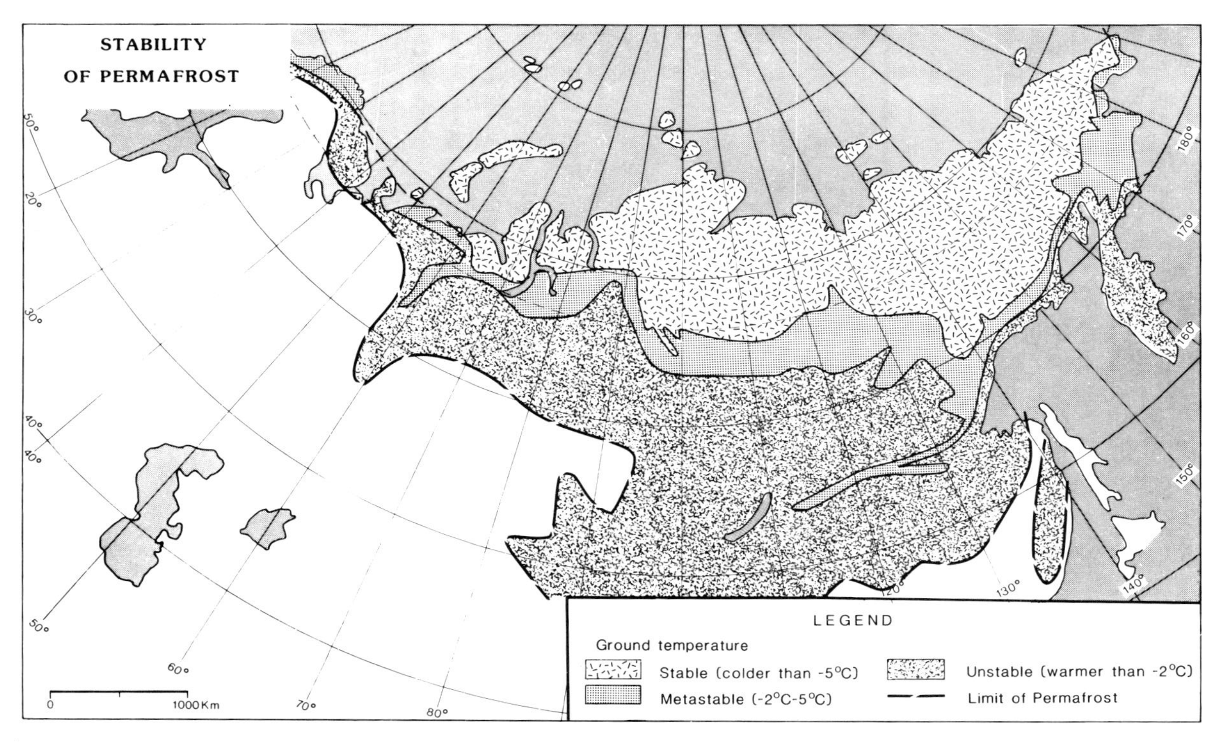

Figure 3.10: Thermal Stability of Permafrost in Asia

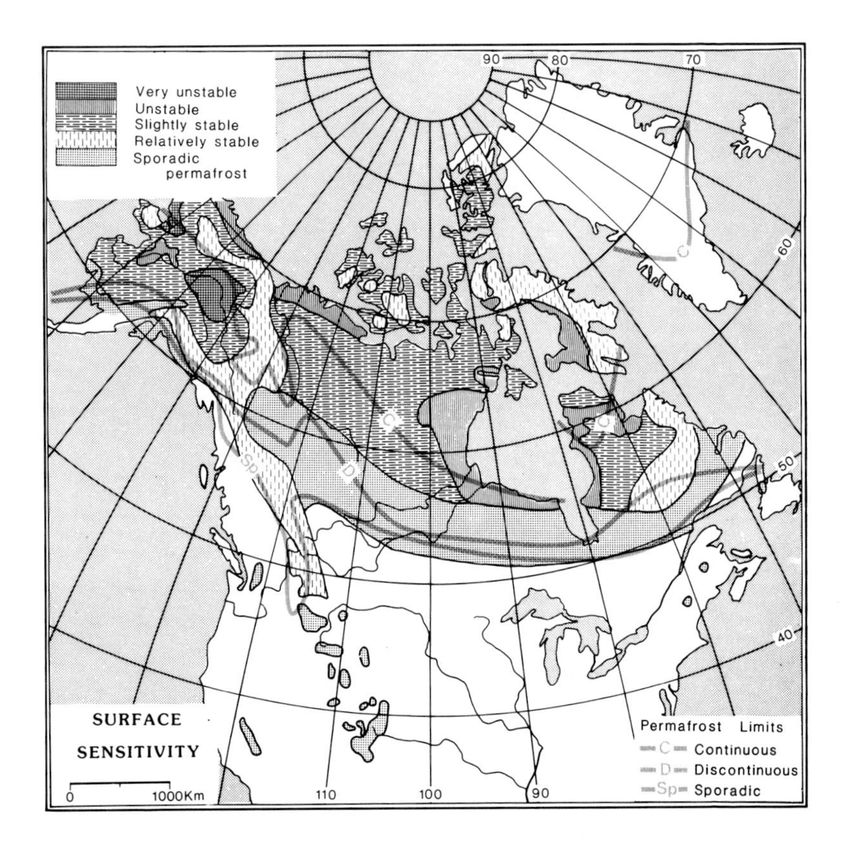

Figure 3.11: Sensitivity of the Ground to Removal of the Soil and Vegetation Cover in North America (after Grave, 1984b).

permafrost areas according to their stability when subjected to environmental changes. Two systems have been developed, one based strictly on the ground temperature regime and the other taking into account the instability when ice melts.

Cheng Guodong (1983) suggested replacing the conventional permafrost distribution maps of China by maps of permafrost temperature. However, Harris (1986) pointed out that these maps serve different purposes. Harris subdivided permafrost into stable (colder than -5°C), metastable (-2° to -5°C) and unstable (permafrost warmer than -2°C) thermal classes. Figures 3.9 and 3.10 show the system applied to North America and Asia. The striking

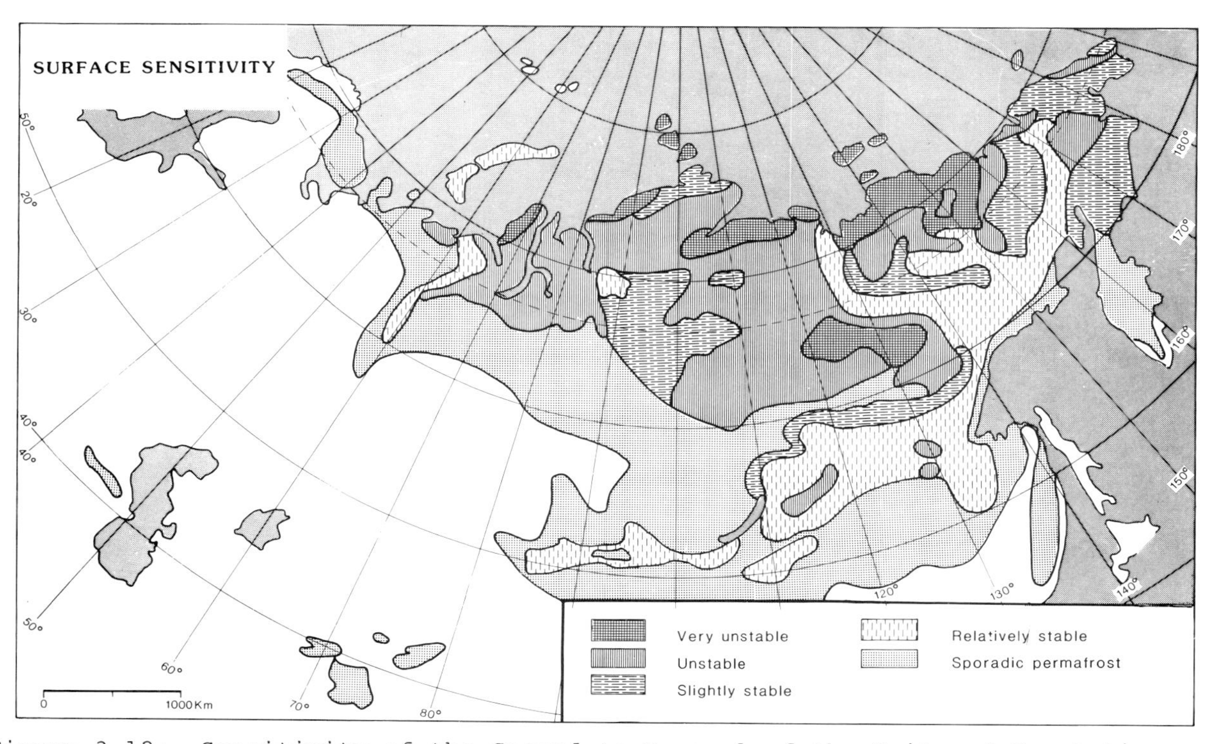

Figure 3.12: Sensitivity of the Ground to Removal of the Soil and Vegetation Cover in Asia (after Grave, 1983).

feature is that thermally stable permafrost is only found around the Arctic basin and in northeast Siberia, while unstable permafrost accounts for at least 40% of the permafrost areas on these two continents. Thus, a warming of air temperatures by 2°C would cause the degradation of permafrost, which may result in the development of thermokarst in over 40% of these permafrost areas. Compared with some recent estimates of the thermal effects of increased carbon dioxide, such a temperature change is quite possible.

Maps of thermal stability disregard the ice content of the ground. Grave (1983, 1984b) has produced maps of the discontinuous and continuous permafrost areas, classifying them according to sensitivity of the surface to removal of the soil and vegetation covers under the existing climate (Figures 3.11 and 3.12). The most unstable areas are generally coastal lowlands and deltas but maps on this scale do not show the variability that may be present within an area. On the whole, the permafrost areas of North America are less susceptible to man-induced thermokarst than those in the USSR, due to the greater incidence of thin soils over bedrock resulting from glacial erosion of most of the area. Brown and Grave (1979a, 1979b) discuss the general principles of protection of the environment during interference by man, while Grave (1984a) discusses environmental protection in permafrost terrain in the USSR; Brewer (1984), French (1984), Hemming (1984), McVee and Tileston (1984) and Webber (1984) discuss aspects of the problems and progress in North America.

EFFECTS OF CLIMATIC CHANGES

McBeath (1984) provides a summary of some of the possible effects of a rise in air temperature in polar regions induced by the increasing carbon dioxide content in the atmosphere. Even if the air temperature remains constant, plant growth will be modified by the changes in carbon dioxide content (Wittwer, 1984), thus altering the local ground thermal regime.

Figures 3.9 and 3.10 can be used to estimate the areas where melting of ground-ice would occur under natural conditions for a given temperature change. The changes would occur progressively until a new equilibrium was reached. An overall temperature increase of 4-5°C for the earth's surface has been postulated, with a two to three-fold greater effect

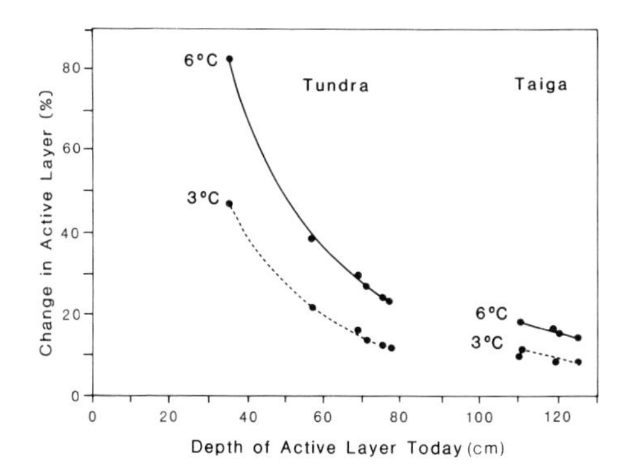

Figure 3.13: Changes in Thickness of the Active Layer Along the Dalton Highway (Fairbanks to Prudhoe Bay, Alaska) as Calculated Using the Stefan Equation and a Positive Change in Mean Annual Air Temperature of Either 3°C or 6°C (data from Goodwin et al., 1984).

in high northern latitudes (Kelly & Gosink, 1984). Such a change would cause the disappearance of the sea ice in much of the Arctic Ocean each summer (Kellogg, 1984).

Goodwin et al. (1984) have examined the theoretical effect of a 3° and 6° warming on the thickness of the active layer using a Stefan equation solution, for stations along the Dalton Highway between Fairbanks and Prudhoe Bay. As seen in Figure 3.13, the results suggest different percentage changes in thickness of the active layer for tundra and taiga vegetation. Maximum percentage change will be in soils with a thin active layer under the tundra. Thus, even in areas where the permafrost persists, there would be substantial changes in the active layer, which have not been allowed for in current engineering designs. In the high Arctic, these would probably exceed the safety margins currently used.

Fortunately, the long-term data from weather stations and ground temperature cables do not indicate a warming trend to date, but the situation obviously merits careful scrutiny.

Plate V: Log House, Irkutsk, USSR, Sinking into the Ground as Thaw Settlement Occurs.

Plate VI: Log House that has been Jacked up so that Gravel can be used to Fill the Void Beneath it (Fairbanks, Alaska).

Chapter Four

FOUNDATIONS IN PERMAFROST REGIONS

Modern construction on permafrost makes use of one or more of a specialized group of construction techniques on thaw-sensitive and frost-susceptible materials. These methods are varied somewhat according to the type of structure involved, the nature of the ground and the locally available materials, but they are basically the same for most construction purposes.

In this chapter, the nature of the problems associated with permafrost will be examined, followed by a description of the methods used for measuring thaw-sensitivity and frost-susceptibility. This is followed by a systematic description of the basic construction techniques used to overcome the problems.

THE PROBLEM

Deformation or damage to structures (Plate II) is caused by seasonal freezing of water and thawing of ice in the soil underlying and adjacent to the structure. Either or both may be involved and with time, the soils change in strength. An even greater problem is the thawing of ground ice due to disturbance of the thermal regime of the permafrost.

The problem during seasonal freezing is that the soil becomes bonded to the foundation and when subsequent heaving occurs due to ice-segregation, the building can be lifted up. If differential movements are serious (as they usually are), the building will be liable to fail, but if there is an even amount of heaving, this is not necessarily a problem unless it exceeds the allowable movements for the structure.

The problem during seasonal thawing is that the ice that has been accumulated in the frozen surface layers melts and takes up less space and the soil loses strength, thus causing thaw settlement. The time at which melting under the building takes place may vary; some materials with high ice contents, such as silts, may flow when thawed and result in additional settlement. Variations in the amount of ice in the soil around the foundation will cause differential settlement in thaw-sensitive soils and will put the foundations under stress, which may cause failure or unacceptable deformation of the structure.

Additional stresses will be produced if progressive thawing of the underlying permafrost takes place with time. These stresses gradually increase with time since the thawing is cumulative and may eventually dwarf the seasonal effects (Fig. 4.1).

From this it will be seen that the greatest problems are usually with thawing around foundations in thaw-sensitive soils.

DESIGN OF FOUNDATIONS FOR STRUCTURES

Detailed site studies are mandatory to determine site conditions and to anticipate the nature of the problems in a specific case. It is essential to understand the stratigraphy and thaw-stability of the soil and underlying permafrost, the ground thermal regime, and the strength and deformation characteristics of the foundation soils. The aim of the study is to produce a design that will avoid failure of the foundation and unacceptable deformations during the service life of the structure.

Thaw-sensitivity depends on the amount and disposition of the ice. If the latter is interstitial and is confined to the pores, the substrate will be thaw-stable. If segregated ice or other ice masses occur outside the pores, the soil will be thaw-sensitive. Moisture contents up to 1600% by weight may occur in silts.

The usual procedures in designing for structures in warmer climates involve determining the foundation design loading, the representative soil strength and deformation characteristics, and then selecting a foundation design that spreads the load in such a way as to prevent soil failure and to avoid unacceptable settlements.

When a heated building is placed on permafrost without adequate precautions being taken, a thaw bulb normally develops beneath it (Figure 4.2, after RIGCDR, 1975, Fig. 5). Zhao Yunlong and Wang Jianfu (1983) and Jumikis (1978) give examples of graphs and

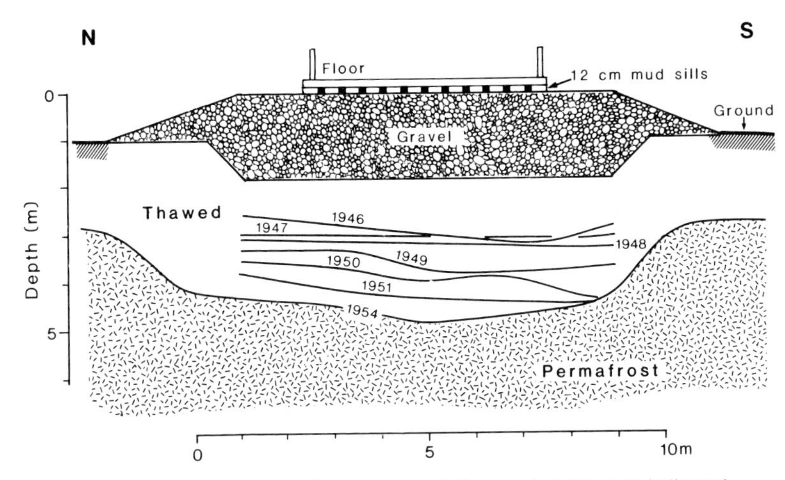

Figure 4.1: Progressive Degradation of Permafrost Beneath the 1.83 m Gravel Pad of Building #4, Alaska Field Station, 4.5 km Northeast of Fairbanks, Alaska (after Lobacz & Quinn, 1966). The active layer was 0.90 m thick prior to construction in 1946 but increased to about 1.75 m after clearing of the trees and shrubs.

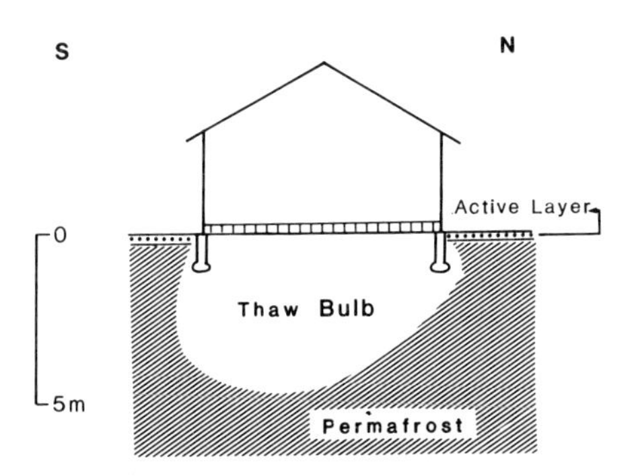

Figure 4.2: Cross-section Under an Experimental Building After Three Years of Use (after RIGCDR, 1975, Fig. 5). Reproduced with the permission of the National Research Council of Canada.

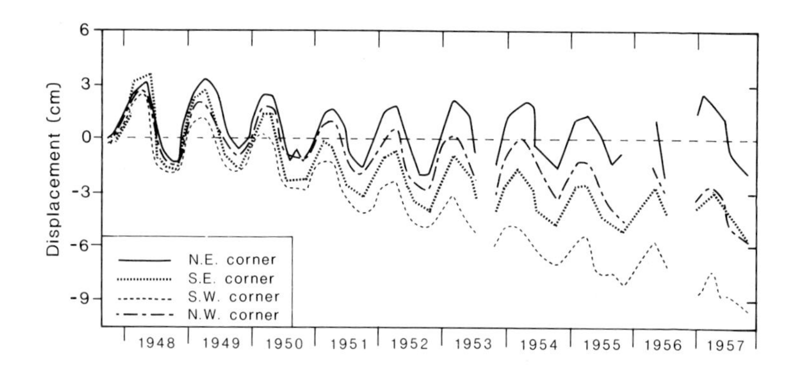

Figure 4.3: Differential Settlement and Heave at the Four Corners of Building #9, Alaska Field Station, 4.5 km North-east of Fairbanks, Alaska (after Lobacz & Quinn, 1966). The foundation consisted of two concrete slabs (upper 15 cm, lower 22.5 cm) separated by 90 cm high concrete piers, with 1.71 m of gravel pad beneath the lower slab. The movements represent (1) seasonal frost heaving and thawing of the active layer (up to 5 cm amplitude) and (2) progressive long-term thawing of the underlying permafrost (up to 10 cm in 10 years).

formulae for calculating the probable thawed depth and temperature disturbance produced. The thaw bulb is asymmetrical due to the differences in solar heating between the north and south sides of the structure (see Lobacz & Quinn, 1966; He, 1983). The aspect modifies the seasonal temperature conditions, the moisture movement, the ice contents, heave and soil stability around the building, resulting in differential seasonal heaving and settlement as shown by the relative movements of the corners of the building (Fig. 4.3). The long-term settlement is progressive and gradual and will continue until a new thermal equilibrium is established some decades later. The exact time of freezing and thawing and the magnitude of the thermal changes will be different at the four corners of the building due to the effects of aspect. On thaw-sensitive soils, this will tend to produce failure of the foundations and cracking of the walls. Even on thaw-stable soils, the building may fail if long-term thawing of permafrost occurs.

Not only must the heat loss into the ground below the building be considered, as in Figure 4.2, but there will also be disturbance of the thermal regime of the surface around the structure as a result of the construction activities (G. H. Johnston, 1981). Once again, this will tend to cause further foundation problems. Utilities, pathways, etc. will add to these thermal disturbances, while the settlement of the building may affect utility joints and connections.

Accordingly the design approach used in permafrost areas starts with the determination of whether the soils are thaw-stable or thaw-sensitive. On thaw-stable substrates, the conventional methods of foundation design used in non-permafrost areas are utilized.

On thaw-sensitive substrates, a decision is made as to whether to use passive methods designed to preserve the ground thermal regime or to use the active methods which accept and control the thaw resulting from the changing thermal regimes to suit the design. In general, the passive methods are used in areas of both continuous and discontinuous permafrost which are relatively stable, whereas the active methods are used primarily in areas of unstable permafrost in the discontinuous permafrost zone. The latter may involve keeping the thermal changes within acceptable limits, or the complete elimination of the permafrost or the unfavourable ice-rich, thaw-sensitive soils prior to construction.

There are at least ten techniques that can be used to promote safe building construction on permafrost. Passive methods involve the use of piles, spread footings, posts and pads, ventilation ducts, artificial refrigeration, rigid structural bases (slabs or rafts), sills or end-bearing piles, caissons, or footings on a deep stable stratum. These will be discussed below.

The two main active methods are the pre-thawing and pre-consolidation of unfavourable materials and the replacement of such substrates by thaw-stable materials. The former is used in areas of sporadic permafrost along road alignments where the right of way is cleared of vegetation several years before construction begins. The zone is then kept bare so that it absorbs more heat from the sun and any permafrost thaws. A similar technique is currently being used on new farm land in northern British Columbia, for example, just west of Fort Nelson. The burning of the vegetation in piles on the site also helps warm the ground. Replacement of the material is commonly practised on peaty deposits, either for highways or for other structures.

Passive methods work well in stable permafrost, or in most coarser substrates with metastable permafrost where the ground temperatures are constantly below -3°C, provided that the ground is kept frozen. This is due to the greater bearing strength of these colder soils due to the greater amounts of moisture in the form of ice, cementing the soil particles together. However, some metastable clays may have a high enough liquid water content to have a low bearing capacity.

Structures are usually designed to be entirely above ground. Basements below ground entail excavation into the permafrost, and when the inevitable thawing occurs around the walls, this results in drainage and settlement problems. The only exception is on dry, dense, thaw-stable materials. In all cases, good drainage away from the foundations is essential to avoid ponding of water and thermal erosion.

For engineering design, conservative values for ground temperatures are used based on mean ground temperatures at depth. Where possible, these are measured for a period of one to two years and a safety margin is allowed for. However, it is known that climate is not constant and there can be considerable climatic variations from year to year and over longer periods of time. Usually, the long-term changes are poorly known and not allowed for. However there are a few temperature strings that have been monitored regularly for long periods, e.g. by G.H. Johnston (1984, personal communication) at Inuvik (since 1955) and at Thompson, Manitoba; by J.R. Mackay in the Mackenzie Delta; and since 1974 in the front ranges of the Rocky Mountains of Alberta and British Columbia (Harris & Brown, 1978, 1982; Harris, 1985).

TYPES OF FOUNDATIONS

Various types of foundations can be used singly or in combination. In the following account, they will be discussed singly as far as possible.

1. Pads

These were one of the original methods of construction used in permafrost regions. They consist of a layer or pad of frost-stable material laid on the surface of the active layer or in a shallow excavation.

Pads used as foundations for roads, airfields, railways, etc. are exposed to cyclical seasonal

Plate VII: Leaking Oil Storage Tanks, Inuvik, Northwest Territories. Heat from the tanks has caused differential thaw-subsidence, resulting in oil leaks where the metal plates meet.

Plate VIII: Building the Ventilated and Insulated Pad for a New Oil Storage Tank, Inuvik, Northwest Territories.

temperature fluctuations. In these cases, the thickness of the pad should exceed the calculated thickness of the new active layer after the structure is completed and the new thermal equilibrium has been reached (see Chapter 5).

Pads used for buildings are subjected to a more or less constant heat source in addition to the cyclical seasonal fluctuations. In theory, the thickness of the pad would have to exceed the calculated thickness of the thaw bulb under the building after the total design life of the structure had elapsed. In practice, this is far too great a thickness of pad without some additional methods being used (see Plates V & VI). It forms a good working surface to reduce frost heave but in addition, an air space, insulation in the floor, a duct system in the pad, etc., will be needed to aid in controlling the size of the thaw bulb to keep it within acceptable limits.

Lobacz and Quinn (1966) demonstrated that there is a substantial and progressive change in the permafrost table beneath pad foundations with time (Fig. 4.1) in areas of unstable, discontinuous permafrost. Unfortunately the building was dismantled in 1954, but it appears likely that thermal equilibrium was being approached after eight years. Where a thin gravel pad is used, the thaw bulb is even more pronounced, even when a 75 cm closed air space is included in the design (He, 1983). Figure 4.4 shows the extreme seasonal thermal regimes beneath a building at the Zhaohui field station in northeast China.

2. Slabs or Rafts

Slabs or rafts are rigid floors, usually made of reinforced concrete. They are commonly laid over a gravel pad of suitable thickness and provide a smooth, uniform base on which to build. The load of the building and its contents are spread more evenly over the entire base of the slab than with sill or pile construction, so that provided it is strong enough, it can carry greater loads. When differential movements of the ground take place, the slab will be tilted, but unless it fails, the walls will not be subjected to the stresses.

Slabs are commonly used in power-houses, warehouses, garages and unheated buildings. Since concrete has relatively high thermal conductivity, it is usually necessary to use additional methods to prevent permafrost degradation. Thus, in building #9 at the Alaska Field Station, Lobacz and Quinn

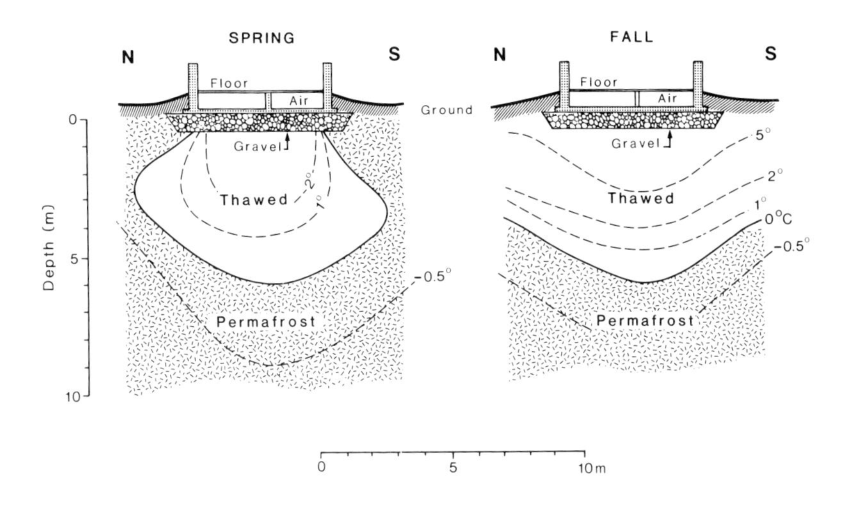

Figure 4.4: Isotherms Below 75 cm Sill Foundation with an Open Air Space Overlying a Thin Gravel Pad at the Zhaohui Field Station in Northeast China (after He, 1983). The original thickness of the active layer was 1.6 m.

(1966) used two slabs separated by 90 cm high piers to allow for air circulation. These were placed on a 1.71 m gravel pad. It will be seen in Figure 4.3 that progressive tilting occurred and was still occurring after ten years. Without the piers, the amount of settlement in the same time period was at least three times as great, although the amount of tilting was the same.

Obviously, the deformation in Figure 4.3 is unacceptable. The most common methods of reducing the thawing to acceptable levels beneath slabs are the use of insulation within the pad and thermal piles, pipes and cold air ducts (Plate VIII). These will be dealt with below. Without these, long-term failure of the structure will normally occur (Plate VII).

3. Sills

These are the simplest form of foundation and consist of beams laid on top of or just below the surface of a compacted pad, or even on the ground surface. Sills are used for small, lightweight, or lightly-loaded buildings and are usually unheated.

They can tolerate some seasonal movement and overall settlement with time. Wood is the most usual material and the log houses formerly used in Siberia and in northern Canada would be examples. Ideally, an air space is left between the pad and floor to provide ventilation and to act as a protection from moisture. The floor must be insulated if the building is to be heated.

In the older log buildings, the walls would be caulked every year to repair the damage from differential movement. When the cabin sank deeply enough for the ground surface to reach the window sills, the building would be dismantled and gravel emplaced under it. Then it would be rebuilt. This is still done in Siberia, but in Alaska, such a building is jacked up and then the gravel is placed beneath it and the services reconnected (Plates V & VI).

In timber-frame construction, the walls are bolted to the sill with extra long bolts to allow for placing packing under the walls as necessary as settlement occurs. Even with the bolts, the buildings may be blown off the sill foundation in strong winds, so that the buildings need to be anchored to the foundation soil.

This type of construction is still used to reduce costs where the buildings are not going to be used throughout the year. However, in general, it has been replaced by other types of shallow foundations for small buildings or, in some cases, deeper foundations may be used.

4. Spread Footings

This is a technique used in areas of continuous permafrost with a thin active layer. Under such conditions, refreezing of the ground is fairly rapid after disturbance, which is a major consideration during construction. Spread footings consist of reinforced concrete, wood or steel footings shaped like an inverted "T", placed so that the base is anchored firmly in the permafrost. The base should be placed at a depth equal to at least twice the eventual thickness of the active layer that will occur during the lifetime of the structure, so that the restraining forces of the permafrost on the footing can overcome the heaving forces developed in the active layer.

The footings may be emplaced in a suitable gravel pad laid on the original gravel surface, often with a layer of insulation above the footings to promote the permafrost rising into the fill around

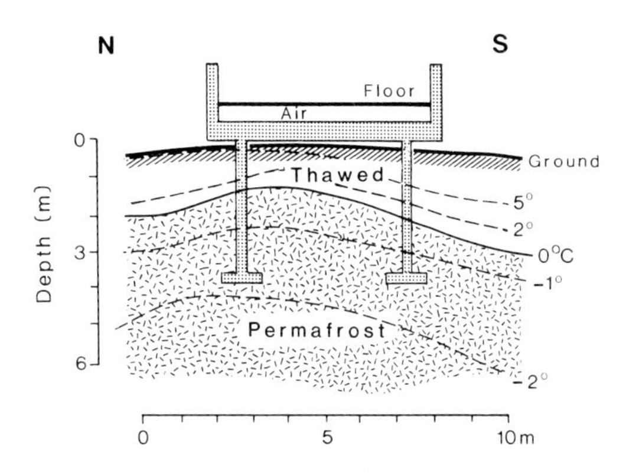

Figure 4.5: Ground Temperatures at Maximum Depth of Thaw Beneath a Pier-Supported Building on Spread Footings, Five Years After Construction at Jingtao Station, Northeast China (He, 1983). The original permafrost table lay at 0.4 to 1.0 m depth. The soils contained 70-80% ice by weight.

the flat base of the footings. The alternative is to excavate a suitable hole in the permafrost and then emplace the footings in a compact, frost-stable granular fill. Sometimes a greased collar is used to protect the footings from frost heaving, and peat or other insulating material may be placed on the ground surface to reduce thawing.

The footings may be at the base of individual piers or long wall-like structures. The latter give more problems if differential settling occurs, whereas shims can be placed on top of piers that show abnormal settlement. To be successful, a clear airspace at least 0.7 m high must be left above ground and adequate insulation must be included in the floor of the building to minimize downward heat loss. The restraining forces must exceed the heaving forces on the piers and footings. The long-term heaving capacity of the ground and associated creep settlement must be allowed for. Excavation and backfilling must be carried out so as to create the minimum ground thermal disturbance. Winter excavation from a snow pad is ideal, but a corduroy overlay of logs may be necessary to protect the ground surface from excessive disturbance during summer construction. Placing the footings in a gravel pad placed on top of the original ground creates the least thermal disturbance. However, it is essential that the piers be left without a load until the ground has completely refrozen around the foundation. The building is only erected when the ground temperatures around the piers and footings have reached a level where adequate strength has been achieved to take the weight. This may mean a delay in construction of one to two years, which is considered poor construction practice. Temporary mechanical refrigration may be used to speed up the freeze-back.

Successful construction of buildings of varying size with this method has been reported by Dickens and Gray (1960), Harding (1962) and He (1983). However, some problems were encountered in the case of the hospital at Kotzebue, Alaska due to very complex foundation conditions near the ocean (Crory, 1978). A combination of very variable material and stratigraphy, very variable salt content in the soil and groundwater, and variable permafrost conditions produced considerable problems. Although the spread footings can be very successful, they tend to be avoided in favour of piles, which are more easily placed in the ground.

Typical ground temperatures at the time of maximum thaw for a pier-supported building at Jingtao Station in northeast China are shown in Figure 4.5. The ground temperatures had stabilized within five years but there were about 6 mm of differential seasonal heave of the foundation. Considering the high ice contents of the permafrost, this is an excellent performance.

5. Piles

Pile foundations are widely used in permafrost areas (Vyalov, 1984; Ding Jinkang, 1984; Ladanyi, 1984). They consist of relatively thin vertical supports with a substantial air space between the base of the building and the ground surface. The vertical supports are embedded in permafrost and can support a very heavy load as long as they are properly installed in a strong frozen bed which is thermally and structurally stable (Plate IX).

He (1983) provides two examples of the thermal regime beneath pile-supported buildings in northeast China (Fig. 4.6). They are comparable in efficiency to pier-supported structures but are more easily emplaced.

Piles may be made of wood, steel H beams, steel pipe or precast reinforced concrete. It is also possible to use piles whose shape varies, e.g. they may be wide at the base to spread the load and thin in the active layer to reduce the uplifting forces

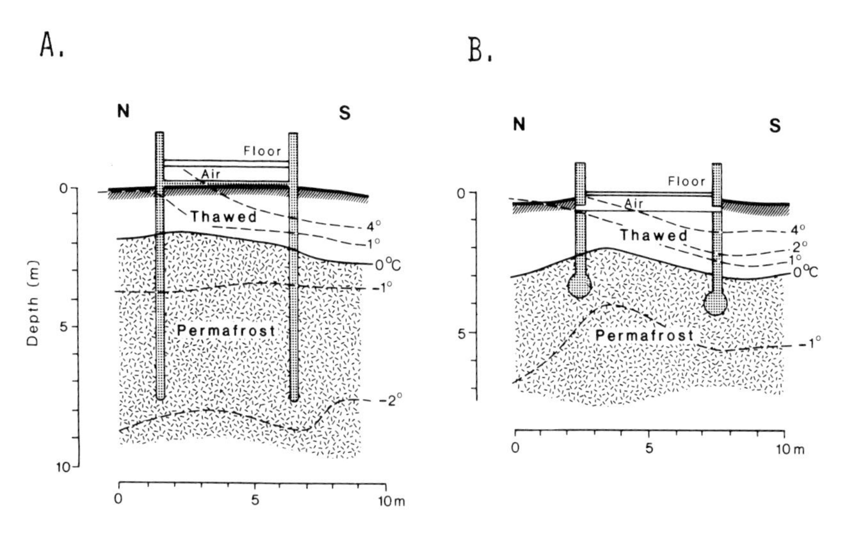

Figure 4.6: Maximum Ground Temperatures Beneath Buildings Constructed in Northeastern China, A, by Drilling Pile Foundations at Jingtao Field Station, and B, by an Explosively Enlarged Pile Foundation at Zhaohui Field Station (He, 1983).

due to seasonal frost heave. Wood is commonly used in forested areas, e.g., north to Arctic Red River along the Mackenzie valley in Canada. The steel H beams will take heavy vertical or lateral stresses and can be driven into warm, plastically frozen fine-grained soils. This minimizes thermal disturbances (Davison et al., 1978) but can only be used where a pile driver is available. Drilling a small pilot hole and filling it with water at above 66°C for 30 minutes in sands and 60 minutes in clays prior to pile driving, greatly speeds up the production rate (Manikian, 1983).

Steel pipes fill drilled holes better than H beams and can be filled with concrete or sand to increase their load capacity, or they can be readily modified to allow for remedial cooling as thermal piles at a later date if necessary. Concrete piles are expensive and liable to crack unless adequate reinforcing is used. They are strong in compression but weak under tension. However, they can be fabricated on site if good clean gravel is available. This is the main type used in the USSR. Concrete piles should not be cast in place, to avoid the effects of heat of hydration and of the low ground temperatures which may inhibit proper curing and

Plate IX: Apartments Built on Piles, Inuvik, Northwest Territories.

Plate X: Wooden Sidewalk on Piles Near Lake Baikal, Siberia.

build-up of strength in the concrete. The scaly outside of poorly cured concrete gives negligible adfreezing of the soil to the resultant pile. Salts may adversely affect the strength of concrete.

The main zone of rot in wooden piles is at the ground surface and in the active layer. Creosote is normally used as a preservative. Much of the strength of piles is obtained by the freezing and friction of frozen ground on the sides of the pile. This is called the adfreeze strength. Creosote and other coatings reduce the adfreeze strength along the treated part of the pile by at least 20% (Aamot, 1966; Parameswaran, 1978) and must be allowed for. This is often advantageous in providing resistance to frost heave.

Piles are usually installed from gravel working pads to protect the ground from damage. The site needs to be adequately drained to avoid ponding of water around the piles and subsequent thermal erosion. Timber corduroy mats can be used as a temporary working surface if no gravel is available. In winter, a snow or ice pad will be satisfactory.

Holes may be produced by steam-thawing, augering, boring or driving into undisturbed frozen ground. Steam jetting was used in Canada until about 1960 (Pihlainen, 1959; Johnston, 1966), but it introduced too much heat into the ground. It is still used in the USSR. The thawed zone around the steam jet often becomes too large in ice-rich and stony soils. This meant that freezeback of the thawed material in the hole (into which the pile was driven) would take a very long time. The holes may be up to 8 m in depth, and will be backfilled with the local soil slurry if they are bailed out. This gives a weaker adfreeze bond in most cases. Freezeback does not work well unless the soils have a mean annual ground temperature lower than -3°C.

Modern drilling or driving of piles makes it possible to install piles as deep as 15 m without seriously modifying the ground thermal regime. Freeze-back then takes only a few days. Augering provides the least thermal disturbances.

The holes should be 10-20 cm wider than the pile to allow for correcting the verticality of the pile and to allow the use of a vibrator to ensure compact emplacement. The slurry is made of free running sands, gravelly sands, silty sands or silts.

The time needed for freezeback of the slurry depends upon the spacing of the piles (Fig. 4.7). This gives some idea of the width of the zone around

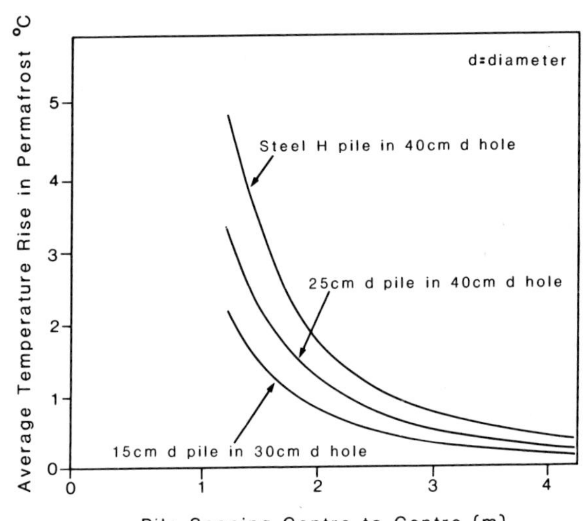

Figure 4.7: Example of the Effect of Slurry and Pile Spacing on Temperature Rise in Adjacent Frozen Ground. Note that these curves are examples only and are for a specific pile size, hole diameter, slurry mix and environmental situation (after. G. H. Johnston, 1981 p. 310). Reproduced with the permission of the National Research Council of Canada.

the pile that is affected by the drilling and back-fill. Driving of piles without any preparation of the pile location gives the most rapid freeze-back but also implies that pile-driving must not stop for more than 30 minutes or the pile becomes frozen in place. Wooden piles tend to splinter.

Winter emplacement is ideal because of the rapidity of back-freezing and the minimal damage to the ground surface. In summer, keeping down thermal disturbance is extremely difficult and freeze-back is greatly delayed.

The structure should be placed at least 60 cm above the ground level, i.e. a clear air space between the floor and the ground surface is required. Placing aluminium reflectors along the perimeter of the structure in summer will tend to reduce summer heating of the air in this space.

The piles must be well anchored in permafrost before the loads are applied. This means that back-freezing must be largely complete before the structure is built. Piles emplaced in late summer, fall or early winter must also resist uplift due to

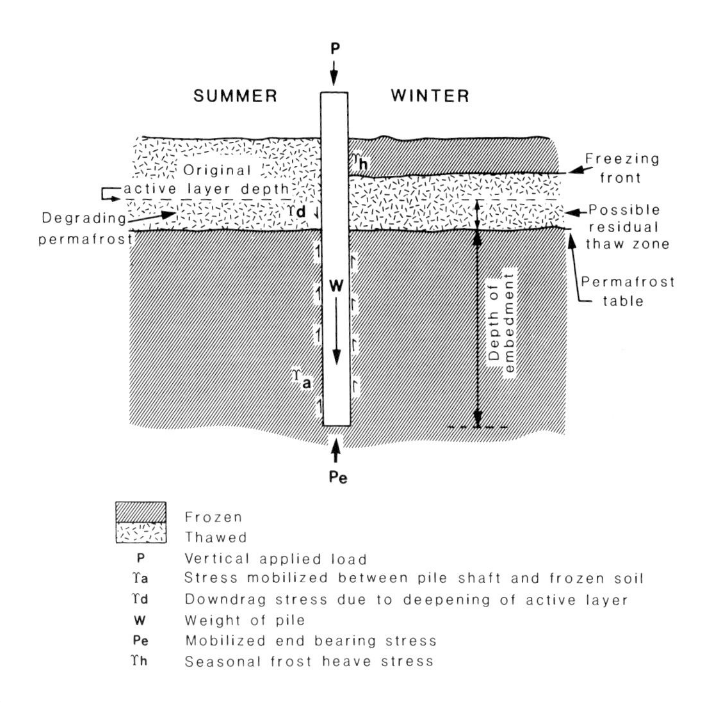

Figure 4.8: Schematic Representation of the Forces Acting on a Pile in Permafrost (after G. H. Johnson, 1981, p. 313). Reproduced with permission of the National Research Council of Canada.

frost heave as the active layer refreezes. This can be overcome by using special measures including ensuring good drainage, the installation of a greased collar between the pile and the active layer and anchoring the base of the pile. This anchoring is usually achieved by embedding the piles deeply in the permafrost zone. Minimum depth below the permafrost table should be twice the thickness of the active layer expected during the life of the structure. The strength is provided by adfreezing of the soil to the pile. Roughening or notching the pile makes relatively little difference in the long-term (Muschell, 1970). However, Andersland and Alwahhab (1983) have demonstrated that lugs on piles placed in a sandy matrix do improve adfreeze strength.

Care needs to be taken to ensure that the wood or concrete piles can stand the stresses produced by heaving and basal restraint in very frost-susceptible soils. Limiting stresses for wooden piles are 4,000 - 5,500 kPa. Steel piles can usually withstand the stresses that are developed.

The collar method is the only successful method in extreme conditions and has been widely used in North America since the building of the Alaska Highway. The main disadvantage is that the resistance of the piles to lateral stresses is reduced considerably by the use of the collar.

The forces acting on the pile are shown diagrammatically in Figure 4.8. The extreme expected thickness of the active layer is a critical factor. Allowance must be made for compaction that may have occurred beneath the work pad. The actual measured thickness of the active layer is increased by 0.3 to 0.6 m to provide a safety margin for design purposes.

Creep of frozen ground is the slow deformation that results from long-term application of a stress to the ground without producing an abrupt failure (Ladanyi, 1972). In permafrost, it is mainly due to creep of pore ice and the migration of unfrozen pore water (Roggensack, 1977; McRoberts et al., 1978). In ice-saturated frozen soils, most creep deformations are distortional with negligible volume change. Where a large unfrozen water content is present or the soils are unsaturated, the deformations can involve volume changes such as consolidation. The settlement of piles in soil can be predicted using the secondary creep law (Nixon & McRoberts, 1976). Figure 4.9 shows the difference between ultimate short-term (less than 24 hours) and long-term sustained strengths of piles. On the whole, strength of the pile increases with decreasing temperature up to a maximum, but the long-term strength is only half that of the short-term strength.

6. Passive Thermal Piles

Thermal piles were developed in 1971 to speed up the freeze-back of piles in discontinuous permafrost and to avoid problems with local thawing. Indeed, they can be used to reduce the ground temperature by about 1°C in areas where the permafrost is unstable, i.e., it has a temperature of about -1°C or warmer, e.g. at Mirny (Porkhaev et al., 1980). Such soils have low bearing capacity and adfreeze strength but this can be improved considerably if the ground can be cooled to -2°C (see Fig. 4.9). Reduction in freeze-back time can speed up construction by several seasons (Essoglou, 1957; Crory, 1965) and thermal piles can effectively counter effects of increased snow accumulation or warming caused by

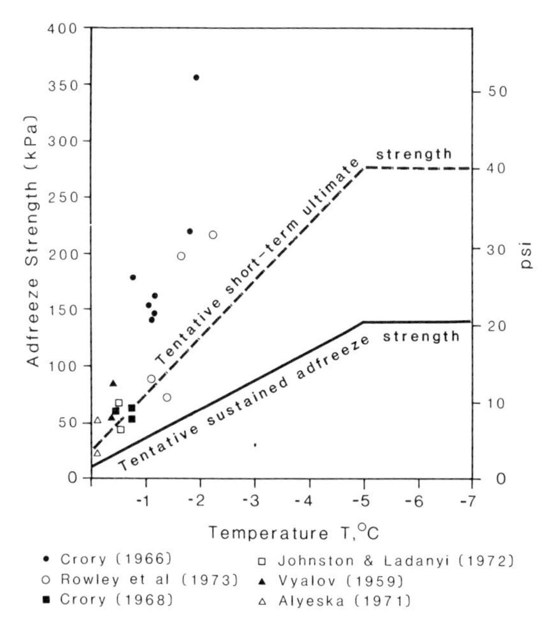

Figure 4.9: Tentative Ultimate Short-term and Sustained Adfreeze Strengths for Wood and Steel Piles in Frozen Clays or Silts with Ice. Note that the relationships are given in metric units (G.H. Johnston, 1981, p. 314). The short-term ultimate strength is only for a load duration of less than one day, e.g. wind or seismic loads. The sustained adfreeze strength does not allow for long-term creep settlements. Reproduced with the permission of the National Research Council of Canada.

construction activity (Miller, 1971; Luscher et al., 1975). They are used for counteracting thawing around foundations of transformers, thawing of margins of palsas below railway lines (Hayley et al., 1983), stabilizing foundations under large buildings where differential settlement is occurring (Hayley, 1982), etc.

Passive thermal piles are natural convection systems utilizing self-powered devices, variously called thermosyphons, thermotubes, convection cells or heat pipes. They have no moving parts, require no external power and only operate when the air is colder than the ground. Individual units of the type used on the Alyeska Pipeline (Fig. 4.10) piles

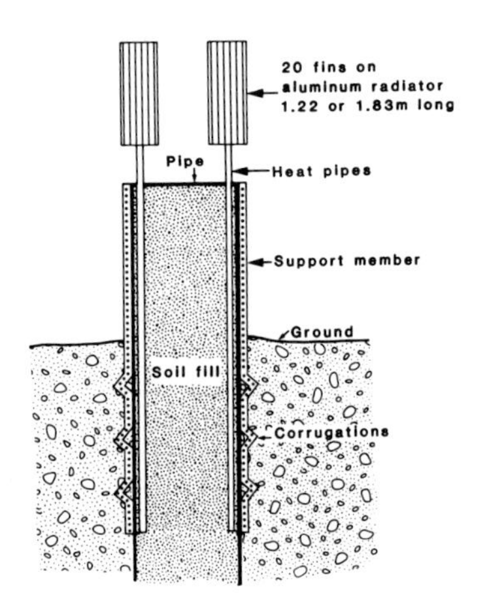

Figure 4.10: Diagram of a Typical Vertical Support Member (USM) Used on the Alyeska Oil Pipeline (not to scale).

cost as little as $250 each. However, they must be monitored regularly to be effective, and for that reason were installed in pairs on the piles on the Alyeska Pipeline (see Chapter 7).

Thermal piles may be either single or two-phase systems (Fig. 4.11). Single-phase devices are usually liquid or air-filled pipes but have so far proved less effective than the two-phase systems (P.R. Johnson, 1971; Babb et al., 1971; Jahns et al., 1973; Reid et al., 1975). Single-phase systems work by convection within the working fluid, transferring heat from the warmer ground upwards to the colder radiator section in winter.

Two-phase systems consist of a sealed tube with the working fluid partly in liquid and partly in gaseous form. Propane, carbon dioxide and ammonia have been used. At low air temperatures, the gas phase condenses on the inner walls of the radiator, releasing heat of vaporisation. This reduces the pressure so that the liquid below boils, absorbing heat from its environs. The condensate descends and the gas rises. This efficient heat transfer system was first suggested by Long (1966) and is ideal for remote locations.

Data from Jahns et al. (1973) show that the performance depends partly on the depth below the surface, partly on the texture of the soil and partly on the length of time the heat pipe is operating.

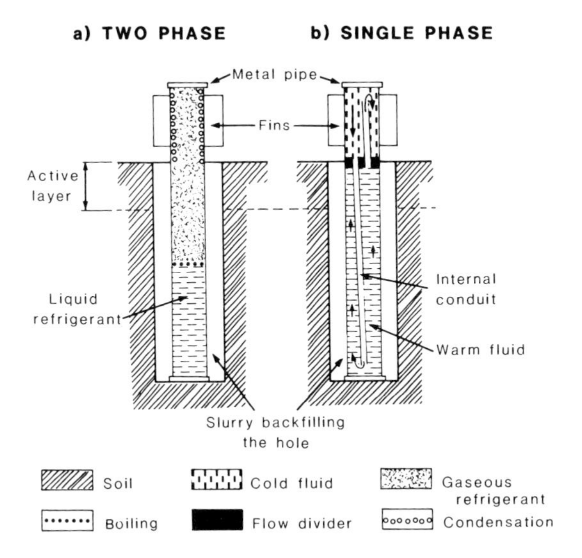

Figure 4.11: Comparison of the Basic Single Phase and Two Phase Thermal Pile Systems (Johnson, 1971).

Cooling is progressive with time and is most effective in clays. Although ground temperatures may be reduced by 4°C around the pile in winter, mean annual ground temperatures are rarely decreased by more than 1°C, due to heat conduction into the ground along the metal pile in summer. Thermal piles can be installed inside hollow steel piles to maintain the frozen condition of the ground.

Most systems used so far were of the two-phase type. The advantage of these is that heat transfer begins when the radiator (Plate XIII) cools to as little as 0.006°C below the temperature of the evaporator; they are therefore well suited to heat removal in the discontinuous permafrost zone. In single liquid systems, heat removal only commences when the temperature differential reaches 3-4°C. In spite of this drawback, Cronin (1983) has shown that the liquid natural convection subgrade cooling system can give comparable heat removal from beneath buildings in ice-rich continuous permafrost, e.g. at Point Barrow, Alaska. Should a liquid natural convection subgrade cooling system start to leak, the whole system will fail, whereas with two-phase units, only the leaking pipe will malfunction. In cold weather, thermal infra-red photography shows up any radiators of two-phase systems that are not working.

7. Artificial Refrigeration

Artificial refrigeration can be used wherever a source of cheap power is available if the situation makes its use an economic proposition. It was first used in the form of freezing coils around newly installed piles to speed up initial freeze-back (Crory, 1973b). Subsequently, they have been used for cooling the foundations of the pumping stations of the Alyeska Pipeline. Greater cooling can be effected than with passive thermal pipes but they are expensive to run on a long-term basis. The moving parts must be maintained and replaced periodically. This type of refrigeration is also used in underground mine portals in North America and in keeping the circulating air in mines cold during summer in the USSR. They are also used for stabilising the foundations of Arctic radar sites where stability of the foundations is essential (Fife, 1960).

8. Ventilation Ducts

The cheapest method of increasing heat dissipation from the foundation pad is by placing ducts or culverts in the pad and opening them whenever the air temperature is below -3°C (Read, 1966; Jahns et al., 1973). At higher temperatures, covers are placed over the ends of the ducts to keep the warm air out. The Balch effect does the work in winter, i.e., cold, dense air displaces the warmer air in the duct by gravity flow.

This technique is used for aircraft hangars, oil tanks (Plates VII and VIII) and maintenance garages at Inuvik, Northwest Territories. These buildings are heated and placed at grade to permit ready access. The floors are usually slabs of reinforced concrete overlying a gravel pad so that they can withstand the heavy loads concentrated at certain points (R.H. Williams, 1959); Sanger, 1969). For larger structures such as hangars, the ducts are connected to vertical stacks with fans, which can aid in the cooling process in winter only. They must be closed off in summer.

It is necessary to orientate the ducts so as to catch the winter winds and to keep them aerodynamically efficient. Such systems must be monitored by thermocouple strings in the pad to ensure that they are operating adequately. Various designs are described by U.S. Army/Air Force (1966), Sanger (1969), Auld et al. (1978), Nixon (1978), Davison and

Lo (1982), Odum (1983) and Zarling et al. (1983). The method has proven very successful where properly used, although failure to close the vents in summer (Davison & Lo, 1982) or allowing snow and water to enter vertical stacks (Tobiasson, 1973) can cause inadequate performance.

9. Insulation

Figures 4.3, 4.4 and 4.5 show that the thickness of a pad foundation in areas of warm unstable and metastable permafrost must be several metres if permafrost degradation is to be kept within acceptable limits. In practice, it is often difficult to obtain an adequate supply of frost-stable material and it is, therefore, advantageous to use an insulating layer. Such layers can easily reduce the required thickness of the pad by 50%.

A natural and abundant insulating material is peat moss which can usually be obtained locally. It has worked satisfactorily in protecting permafrost below centres of road beds on a test site in Alaska (Berg & Aitken, 1973; Clark & Simoni, 1976; McHattie & Esch, 1983) and has been used in Norway under roads and railways to control frost heaving in non-permafrost areas since 1903 (Skaven-Haug, 1959; Solbraa, 1971). It undergoes considerable compaction, but this can be overcome by using compacted blocks. Wood chips and bark serve the same purpose. The frozen thermal conductivity of peat is twice that of unfrozen peat so it performs best over the warm permafrost of the discontinuous permafrost zones. However, the mining of suitable quantities of peat can create aesthetic and environmental problems.

The development of artificial insulating materials has resulted in experimentation in their use in foundations. The insulation should have high compressive strength and not absorb moisture. Rigid board or foamed-in-place polyurethane can absorb water and readily loses its value as an insulator. Expanded polystyrene foam insulation boards are commonly used, and most insulating materials perform better if placed above, below or within polyethylene sheets to reduce the movement of moisture. The exact design depends on the situation. Shifting of the boards due to creep or shift in the surrounding material will open cracks in the insulating layer that greatly reduce its efficiency. This can be largely overcome by having at least two layers of boards that are staggered so that they overlap one another (Gandahl, 1979).

Insulation is especially useful in suppressing "edge-effects" such as along the margins of roadways (Chapter 5), around tower foundations and along the outside of retaining walls (G. H. Johnston, 1981). It is used extensively to reduce the amounts of fill required beneath roads and airfield runways and taxiways, as well as under heated oil storage tanks, etc. (Davison & Lo, 1982) and can be used in stabilization of landslides (Pufahl & Morgenstern, 1979). Nixon (1973) gives a formula for calculating the thickness of insulation needed beneath a building to reduce heating of the ground to acceptable levels.

CHOICE OF METHOD

Usually several of these methods are used together to minimize the movement of the foundation, the thawing and settlement of the ground and the construction costs. Piles have become the most common form of foundation for the lighter buildings and structures, e.g., houses, apartment blocks (Plate IX), schools, sidewalks (Plate X) and utilidors (Plates XIX & XX). Pads and slabs with insulation and ventilation are commonly used for heavier loads which must be accessible.

The same techniques are used in various combinations for other structures such as roads, railways and airfields, as will be described in the next two chapters.

Chapter Five

ROADS AND RAILWAYS

These are essential for the transportation of people, supplies, and goods from one settlement to another. However, permafrost regions have special problems which are only now being adequately dealt with (Esch, 1984). The "state of the art" is demonstrated by the current road and rail networks in North America (Fig. 5.1) and Siberia (Figs. 5.2 & 5.3).

In North America (Fig. 5.1), the year-round road network is well established south of the permafrost areas. It not only crosses the agricultural areas but extends throughout the mountains and onto the Canadian Shield. There it links settlements based on mining, forestry, or the generation of hydro-electric power. However, the road network extends into the zone of continuous permafrost in northern Yukon Territory, western Northwest Territories and in Alaska. The Dempster Highway was completed in 1979, and links the Klondike Highway near Dawson City, (DC) with Inuvik (I). It permits the year-round supply of drilling rigs in the Beaufort Sea, which was not possible before. In Alaska, the Dalton Highway parallels the Alyeska Pipeline from Prudhoe Bay (PB), south to Fairbanks (F). This was built to support the construction and maintenance of the pipeline, but is currently the primary supply route for oil-field support activities.

The railway network also lies mainly south of the permafrost zone, but in Manitoba and Quebec, branch lines have been built across the permafrost to permit the mining of rich ores at Lynn Lake (LL) and Schefferville (S) and to link the port of Churchill with the south. The White Pass and Yukon railway used to provide a service between Skagway and Whitehorse. A winter road network links settlements along and to the east of the Mackenzie valley.

Plate XI: Slipping of the Margin of the Road Embankment Along the Alaska Highway, East of Tok, Alaska, in May 1980(see also Fig. 5.6).

Plate XII: Gold Stream Bridge, Site of the Measurements Shown in Fig. 5.9 (view from the west).

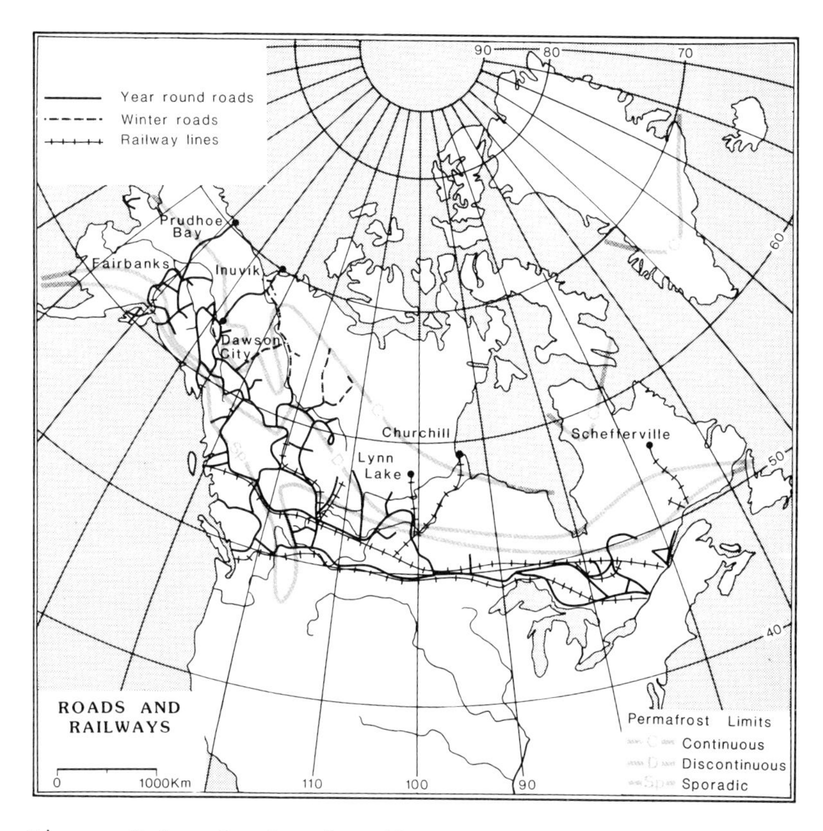

Figure 5.1: Road and Railway Network in Northern North America in Relation to the Permafrost Zones.

In the USSR, the eastwards advance of settlement was along the "Tract", a ribbon of cleared land nearly 100 m wide, stretching from Moscow to Irkutsk. It was first followed by exiles sent east to work in the mines and forests of Siberia in the seventeenth century. In 1753 capital punishment was replaced by permanent exile to Siberia and this same punishment was soon used for a variety of other crimes. Thus a large number of Polish prisoners were sent to Irkutsk in the 1870s but the bulk of the areas settled were south of the permafrost zone. The same system was used after the Revolution until the end of the Stalinist era when the Kremlin sought to recruit young settlers by the lure of high wages and bonuses. In the late nineteenth century,

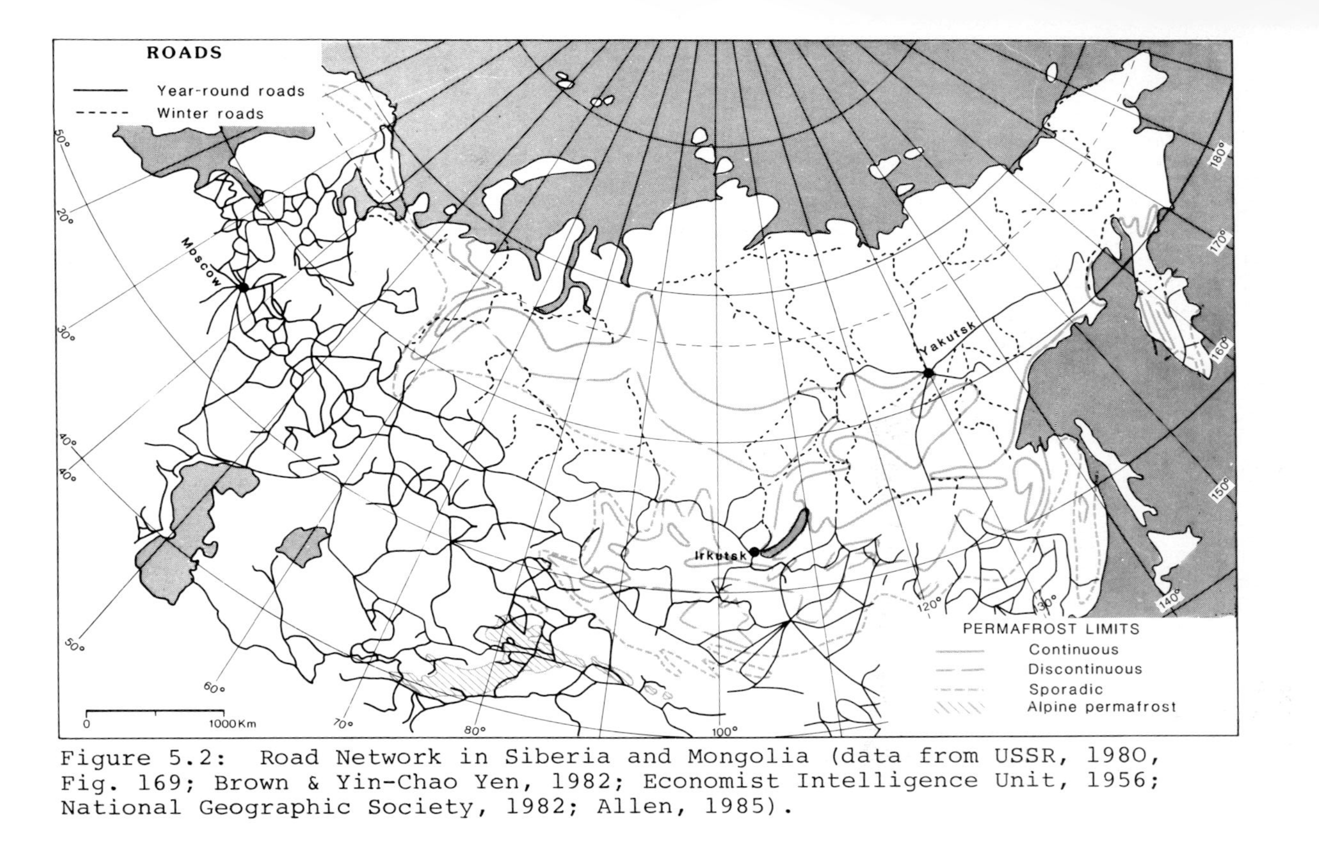

Figure 5.2: Road Network in Siberia and Mongolia (data from USSR, 1980, Fig. 169; Brown & Yin-Chao Yen, 1982; Economist Intelligence Unit, 1956; National Geographic Society, 1982; Allen, 1985).

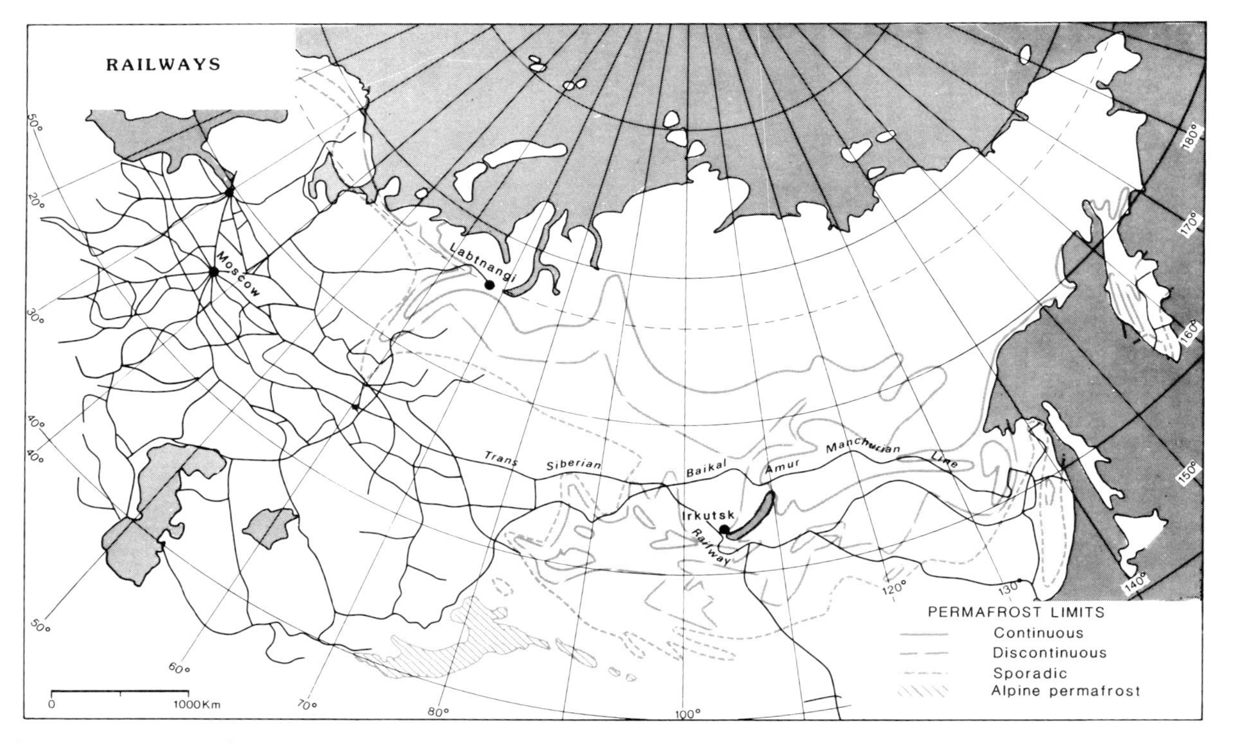

Figure 5.3: Railway Network in Siberia and Mongolia (data from USSR, 1980, Fig. 168; Brown & Yin-Chao Yen, 1982; Economist Intelligence Unit, 1956).

the Trans-Siberian Railway was built to move lumber, etc. westwards.

Figures 5.2 and 5.3 show the present state of the road and rail network based on USSR (1980), J. Brown and Yin-Chao Yen (1982), Economist Intelligence Unit (1956) and the National Geographic Society (1982a). It will be seen that the year-round roads are only just being extended into the zone of continuous permafrost in the region centred on Yakutsk (Y). The rest of the roads in that area appear to be winter roads. Similarly, the railways only cross the zone of continuous permafrost along the Baikal-Amur-Manchuria (B.A.M.) line north and east of Irkutsk (I) and to Labtnangi (L) north-east of the Ural Mountains. The former is based on the military needs for a railway further from the frontier with China, while the Labtnangi line serves important mines and also the port of Salekhard. A similar pattern is seen in China.

THE PROBLEM

There have been numerous studies of the effects of wheeled and tracked vehicles on permafrost terrain. Even a single pass of a vehicle in summer can result in compaction, killing or direct removal of the vegetation and exposure of the mineral soil (J.R. Radforth, 1972, 1973; J. Brown et al., 1969). This results in a modified thermal regime and the development of thermokarst, unless the soil is ice-free (Haag & Bliss, 1974b). Even destruction of the trees changes the form of the heat-receiving zone from a three-dimensional to a more two-dimensional surface, resulting in thawing (Haag & Bliss, 1974a). The trees act as wind pumps that remove large amounts of water from the ground, so their removal usually results in higher water tables and more heat storage in the active layer. Wet peat moss has a much higher thermal conductivity than dry peat, so the active layer tends to deepen in spite of increased evaporative cooling. In soils with high ice contents, this can cause long, narrow strips of water to form where the disturbance took place the previous year, and these tend to enlarge with time until a drainage way is intersected.

Climate affects the thaw-sensitivity of the ground, the permafrost areas with the lowest ground temperatures tending to be least prone to damage. The most easily damaged terrains are those with ground temperatures above -2°C in the zones of discontinuous and sporadic permafrost. Dry soils are normally stable, but the tendency to form thermokarst ponds and lakes increases with increasing ground-ice content.

The consequence of these stability relations is that the zones of discontinuous and sporadic permafrost present an unstable foundation for road and rail construction. The result is the abrupt termination of most all-weather transportation routes at the margins of the permafrost region. However, since 1960, construction techniques have improved to the point where construction on continuous permafrost is possible, although expensive. This is why an all-weather road network is being built centred on Yakutsk and why the B.A.M. line crosses onto continuous permafrost as soon as possible and stays there until almost at the east coast in Manchuria.

WINTER ROADS

The simplest form of transportation route across permafrost terrain is the winter road. It consists of a temporary or permanent winter route across the permafrost terrain and ranges from a simple trail across snow-covered frozen ground or the surface of a frozen river to specially surfaced and compacted snow, ice or ice-water mixtures along a suitable route. In both the Soviet Union and Canada, vast tracts are crossed in this way to supply outlying settlements with bulky, low-value goods and essential supplies.

Their main attribute is that they require no special bed but to be successful on land, the less sophisticated the road construction, the more specialized must be the vehicle using them. Tyre pressure is the critical limitation. Adam (1978) divides winter roads into four main kinds, viz.: winter trails, snow roads, ice roads, and snow-water mixture roads.

Winter Trails

These are simply relatively unimproved routes across the permafrost terrain that are used by specialized winter vehicles. Perennial winter trails are used year after year and usually have a cleared right of way through the bush. Temporary winter trails are used for one or two winters and then abandoned, e.g. after a seismic survey or on completion of a logging operation.

The ground must be frozen to a depth of at least 20 cm to avoid melting and there must be a minimum of 10 cm of snow on the surface to protect the vegetation. The date of achievement of these conditions varies from early October in the Arctic Islands to January on the discontinuous permafrost near Yellowknife. Forest is cleared by a bulldozer equipped with mushroom shoes (wedges fitted to the base of the blade) to avoid scalping the tops of mounds and tussocks.

Only tracked vehicles or those with balloon tyres and low bearing pressure can be used. Trailers and other heavy objects must be fitted on long skis to minimize surface disturbance. The Russian "troika" or horse-drawn sleigh is ideal. Even then, some damage may occur, and successful winter trails can usually be identified readily on aerial photographs taken in summer. Because of their low cost and ease of advancement into a new area, winter trails made up 90% of all the winter roads built in Canada in 1978.

Snow Roads

These are winter roads produced by modifying the snow cover along a perennial right-of-way. It appears as a linear clearing running across the landscape in summer. In winter, it is a highway requiring regular maintenance, the type of which is decided by the nature of the vehicles using it.

Compacted snow roads are roads where the snow is piled on the right-of-way and compacted. Minimum acceptable snow density is 0.4 g/cc, although values close to 0.6 g/cc are preferred. The latter is close to the maximum density obtainable by rolling and dragging. This packing and consolidation results in a smooth, compact surface capable of supporting heavier vehicles with tyre pressures up to 550-620 kPa. The built-up grade reduces drifting of snow, and these improvements permit higher speeds and greater loads.

Processed snow roads are roads where snow is blown onto the right-of-way and then compacted. This process breaks up the snow crystals and allows them to recrystallize in a denser form with less air space. In spite of this, they cannot take the weight of the heavy trucks used on modern paved highways for any length of time.

Ice-capped roads are a refinement of processed roads where the surface has been sprinkled with 2.5 cm of water. This produces a hard cap which wears better for medium weight trucks but quickly becomes rutted by heavy trucks.

The final type of snow road is the manufactured snow road found in areas of low snowfall where it is necessary to link the two ends of a snow road together across a region. They are rare and expensive but can be justified in the case of moving drilling rigs.

Water for ice-capped and manufactured snow roads is preferably obtained from lakes and not from rivers. Removal from rivers may cause the death of the fish population in the streams.

Ice Roads

These are made of ice and come in several forms. Solid ice roads consist of water sprinkled onto the ground surface in successive sheets until it forms a covering over all the depressions forming a smooth surface. It is strong, smooth and fast, but is very costly. It is mainly used in areas of low snowfall.

Aggregate ice roads are a relatively new type of experimental road built of chunks or aggregates of ice chipped from frozen water surfaces and then dumped on the roadway. It takes up to 10 cm of water to bind the aggregates together into a smooth mass.

Finally, there are the "whole river ice roads" which simply use the cleared surface of a frozen river. The USSR has many wide north-flowing rivers without rapids or waterfalls. These provide good routes for transportation by hydrofoil, hovercraft or freighter in summer, and by whole river ice roads in winter. In Canada, waterfalls and rapids and the smaller size of most streams make their use unattractive. The main problems that can occur are icings on rivers and rough surfaces caused by ice jams. However, winter lake ice roads can be built on lakes such as Lake Athabasca by clearing a 60 m right-of-way over the lake ice. The main limitations are the wind-driven pressure ridges that form near the shore and large cracks that occur in the surface of the lake ice.

Ice roads form a much more durable surface than snow roads and permit heavier vehicles to be used. Road speeds can be much higher without breaking up the road. Ice-capped snow roads and ice roads can withstand the traffic and loads to be expected during possible pipeline construction (Adam & Hernandez, 1977). Whole river and lake ice roads can only be used after the ice has thickened sufficiently.

Snow and Water Mixtures

These are roads where the grader pushes the snow into windrows, which are then bowsed with water. The grader quickly spreads the mixture into a smooth surface as it would with hot asphalt. On freezing, it forms an excellent surface for heavy trucks, that can be used for several months without appreciable damage to the underlying terrain.

Ice Bridges

In many locations in permafrost areas, summer crossings of major rivers by conventional roads use ferries. This avoids the problems associated with building bridge footings on permafrost. Examples include the Yukon river at Dawson City, the Peel river at Fort MacPherson and the Mackenzie river at Fort Providence and Arctic Red River. In the fall, winter and spring, the river ice prevents this. As soon as conditions permit during the fall, water is flooded onto the surface of the river ice at a suitable crossing place. This freezes and artificially thickens the river ice until it is strong enough to bear the weight of trucks. The "ice bridge" can then be used until the period of spring break-up of the river ice.

ENVIRONMENTAL EFFECTS OF WINTER ROADS

Until about 1973, there were no regulations regarding the reduction of the environmental effects of winter roads in Canada. However, with the environmental impact assessment of the proposed Mackenzie valley pipeline, a series of studies established the basic rules for minimizing the impact of winter roads on the environment. It was found that their effect need not be great even though a significant proportion of the vegetation may be killed (Younkin & Hettinger, 1978), provided that the underlying peat layer is left as intact as possible. To choose the type of winter road that will produce the least impact, it is necessary to know the type of vehicular use and the susceptibility of the underlying material to a change in thermal exchange rate.

The disturbance itself takes three basic forms, viz: change in terrain morphology, change in snow cover, and a change in the vegetation and peat cover. The usual changes in terrain morphology take the form of chopping off the tops of hummocks and tussocks, while the depressions between the hummocks may widen due to the pressure of compaction.

Ruts may develop in areas such as the eastern Arctic Islands due to sparse vegetation and low seasonal precipitation. However, these properties and the low mean annual temperatures also combine to inhibit runoff and thermokarst erosion (J. R. Radforth, 1972), but the ruts will remain for a long time. Under these conditions, the roads should avoid ice-rich or poorly-drained terrain and run straight up and down the slopes, although this will still cause increased erosion and gulleying.

Winter roads alter the snow thickness and hence the insulating properties of the surface. Compacted snow roads may have 1-2 m of snow which will not melt until long after the snow cover on the rest of the area. This delays thawing and warming of the underlying ground (Kerfoot, 1972; Mackay, 1970). The extra moisture may increase the thermal conductivity of the ground for much of the summer, in spite of evaporative cooling. However, winter trails have little compaction of the snow cover and hence little difference in melt-out time of the snow on the right-of-way. Ice roads take the longest time to melt due to their greater compaction.

Vegetation disturbance is the most pronounced effect. Removal of trees increases the ground temperature by concentrating the thermal exchange zone at the ground surface (Haag & Bliss, 1974a). Coniferous trees give year-round shading and about 1°C lower temperatures than deciduous trees. Trees act as heat radiators in winter. Conifers intercept much of the snow, thus reducing insulation and so increasing winter cooling. Loss of trees also usually results in increased snow depth, giving better insulation of the ground, unless it is blown away. Removal of trees also reduces evapotranspiration and causes higher soil moisture contents (Haag & Bliss, 1974b).

The disturbance of the surface layer tends to loosen and uproot plants (Lambert, 1972), but since the majority of the shrub and heath species are rooted to the sides of the hummocks, they have a good chance of survival. Snow packs into the depressions and helps to protect them.

The peat layer is probably the key to the degree of damage. It thins northward from the southern boreal forest to the tundra. It has high water content and high specific heat but low thermal conductivity compared with mineral soil. This results in more heat being used in melting of ice in spring, retarding heat input into the underlying ground. After the ice melts, the ground starts thawing downwards and the peat dries out, resulting in a layer of low thermal conductivity retarding further heat penetration.

If the peat is partially removed this alters the pattern, increasing depths of thaw and resulting ground temperatures. If mineral soil is exposed, this alters the albedo and results in increased heating.

Compaction of the peat may occur if tyre pressures are too high. This results in increased bulk density in the surface layers and thus increased thermal conductivity. Where the snow cover is too thin, sublimation of the ice may cause voids to develop in the peat. Clearing by hand gives less disturbance of the peat than clearing using a bulldozer.

If the peat and soil are very ice-rich, the thickness of the active layer may double, resulting in significant thermokarst and subsidence of the ground surface. This tends to promote ponding of water. However, if the peat and soil only contain pore ice, the melting results in no more than minor subsidence and thermokarst is avoided (Mackay, 1970).

The amount of subsidence in ice-rich soils can be estimated by measuring the ice content at the permafrost table and estimating the equilibrium thickness of the active layer following thermal disturbance. Even this subsidence for winter roads is considerably less than in the case of summer disturbance (Bliss & Wein, 1972). The subsidence is permanent and irreversible in the short-term, but after revegetation, the active layer will thin and the permafrost will slowly aggrade and accumulate segregated ice lenses, thus raising the surface (Mackay, 1970).

ALL-WEATHER ROADS AND RAILWAYS

These can be dealt with together since the basic problems to be overcome are the same. Both require stable linear rights-of-way extending for great distances and crossing obstacles such as drainage channels and rivers. Both need flat or gently inclined smooth surfaces with the minimum of cut-and-fill and as few drainage problems as possible. The bed of the road or railway will be subjected to vibrations by the traffic, yet it must not fail.

Route Selection

Probably one of the key factors determining the types of problems that will be encountered is the route selection. Aerial photographs can be very useful in selecting the route that is likely to have the fewest problems (Mollard & Pihlainen, 1966). It is essential to determine the surficial geology, geomorphology and hydrology of the area to be investigated, and the person making the decisions must have a clear understanding of the engineering and terrain problems that are characteristic of the various vegetation features in permafrost areas. Detailed drilling and geophysical studies will be needed along the actual route prior to construction.

Linell and Johnston (1973) summarize the main factors to be considered in the construction of road or railway embankments in permafrost regions as being frost action (heave and thaw weakening), thaw settlement, drainage, contraction cracking of pavements and embankments, and stabilisation of embankments and slopes. G. H. Johnston (1983), Goodrich (1983), Huculak et al. (1978), Reid (1974) and Quong (1971) give Canadian examples, while Esch (1983), Lobacz and Eff (1978), Murfitt et al. (1976), Berg (1974) and Hennion and Lobacz (1973) discuss current American experience in design and performance of road embankments. Peretrukhin and Potatueva (1983), Phukan (1983), and Popov et al. (1980) provide an insight into Russian experience.

In general, it is advantageous to locate the road or railway on the crest of a drainage divide, if the topography is suitable, e.g. along the Dempster Highway between Eagle Plains Lodge and the Ogilvie River on the Eagle Plains. This maximizes drainage and minimizes the number of river crossings, culverts, embankments, and icings. When a road or railway has to be built on lower land crossing the drainage from higher land, e.g. the Dempster Highway west of the Richardson Mountains, or where it is built along a valley side in mountains, the problems due to poor drainage, frequent river crossings, culverts, embankments and icings become particularly acute. The availability of suitable frost-stable construction material is often also critical. These problems will be dealt with in turn below.

Railways must be aligned so as to minimize gradients and bends in the track. Since most are primarily designed to carry freight, the gradient should be under 1%. Curves need to be designed to be less than 4% in order to make the most efficient use of tractive power. Obviously, this may not be possible in mountains. In other respects, the factors affecting choice of alignment are the same as for roads.

Source Materials for Pads

Even a 1.6 m pad such as is used on the North Slope of Alaska involves the use of tremendous quantities of fill (23,755 m^3/km). About 65 m^3/km of gravel or crushed rock are needed annually for the maintenance of a 2.75 m wide road.

The main materials used are gravel, crushed rock and clay binder. Crushed rock is quite satisfactory for the main body of the pad but it causes tremendous problems with punctures as a road surface. River gravel has rounded pebbles that make an ideal surfacing provided that a small amount of clay binder is added. The gravel and binder are applied alternatively to produce the final surfacing.

Borrow pits often represent the worst kind of environmental disturbance unless they create over-wintering conditions of deep water favourable for certain species. Stream gravels are best skimmed from braided streams, care being taken to maintain the contours of channel bars. If possible, the mining should be carried out in winter when the flow regime is minimal, and it should be kept away from the actual stream bed carrying the winter flow since winter siltation can adversely affect egg development on gravels.

Where deep mining occurs, it should be in a dyked area and produce a deep hole. If this hole floods, it may be stocked with fish after mining is completed if the water is clean. Fish need pools of water deeper than 3 m for over-wintering, while wildfowl like shallow vegetated margins. Good depth diversity helps, as does an outlet to another pit. Particular attention is necessary to avoid blocking small streams in winter since some species such as grayling use these to spawn in spring.

Rock can be obtained from quarries which should be situated as unobtrusively as possible, e.g. behind hogbacks along the west side of the Richardson mountains. If the rock is blasted loose from a quarry, extra explosives will be needed to do the work because of the elasticity of the ice contained in the rock (see Chapter 8). Talus slopes can be mined with care, although there is always a risk of collapse of a substantial amount of loose rock onto the equipment. Rock can also be scraped off in layers from shallow but extensive borrow pits after the sun has thawed the surface layers in summer. Both these last two techniques are aesthetically less pleasing but much cheaper.

The rock should be of a kind that does not disintegrate and produce silt particles, if possible, although it is claimed by Sørensen (1935) and Taber (1953) that most frost weathering produces silt-sized material.

Clay for the binding fraction must be lacking silt content and is usually obtained from soft marine shales and clays. In Canada, it is spread on the road alternately with the coarse aggregate and mixed in by grading and regrading as the ruts left by passing vehicles are removed. Borrow pits (open and extensive) or quarries can be used as the source. Small quantities are mixed in with the coarse aggregate at most stages of construction or maintenance, unless a particularly free-draining subgrade is required. Clay binders are not used in Alaska.

CONSTRUCTION METHODS

General Principles

The weights of the loads, the vibration and the shape of roads and railways make it desirable to use the pad method of construction with suitable modifications. Correctly installed, the pad will cause the permafrost table to rise somewhat into the pad (Fig. 5.4), in very cold permafrost, although the active layer in a bare pad will normally be much thicker than in the undisturbed soil. The thickness of the pad may be reduced as desired by the incorporation of sufficient insulation. The insulation usually consists of expanded polystyrene foam insulation board laid in at least two layers with staggered joins so that creep in the pad does not open up large uninsulated gaps. Alternative insulating materials which have been used include peat, wood chips and sulphur foams (see Chapter 4). Methods of calculating the required thicknesses of pads and insulation are summarized by G. H. Johnston (1981).

The pad forms an embankment which tends to blow free of snow in winter. This aids in snow removal, and with the latter, ensures greater penetration of winter cold, although the snow accumulated along the shoulders causes warmer ground temperatures. In summer, there is increased heating of the bare ground. This tends to cause problems with thaw subsidence, particularly along the margins of the embankment.

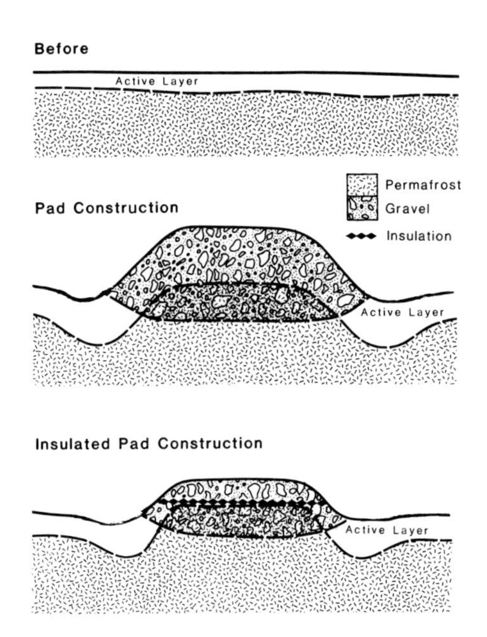

Figure 5.4: Effects of Pad Construction and Insulated Pad Construction on the Permafrost Table Beneath a Road. Note the reduction in volume of gravel needed with incorporation of insulation.

Complete protection from thaw subsidence is most nearly reached in the zone of continuous permafrost where cold, stable permafrost is found. The active layer is thin and so the necessary thickness of pad to provide good protection is more easily achieved. However, Esch (1983) has shown that complete protection cannot be achieved in warm, unstable permafrost (>-2°C) using the methods presently available.

Partial protection is the usual approach. In the zone of sporadic permafrost, the active layer averages 5-6 m but may reach a thickness of 15 m (Harris & Brown, 1978), so that reasonably complete protection would require a pad thicker than 5 m or an insulated pad more than 2 m thick. In practice, this is too expensive so that it is necessary to reach a compromise. Partial protection involves reduction of heave and permafrost degradation to acceptable limits. Thermistor strings have been placed beneath the road at convenient study

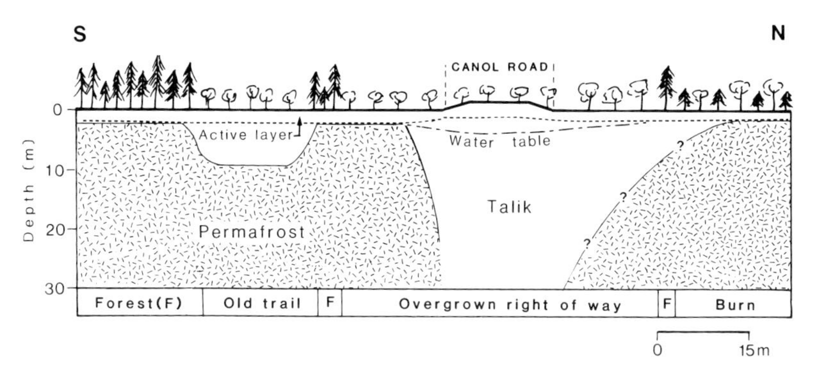

Figure 5.5: Section Across the Canol Road Right-of-Way at Heart Lake, Northwest Territories, in 1972/3 Showing the Effects of Minimal Pad Construction on the Permafrost Table After 20 Years (after Kurfurst & Van Dine, 1973). The road was abandoned within two years of construction and the forest fire north of the road took place in 1969.

locations to monitor changes and test sections built using different construction methods are also studied and compared (see G. H. Johnston, 1983; Goodrich, 1983). Should major repairs prove necessary, the most economical but effective design can be used.

Minimal protection is used for secondary structures such as roads carrying occasional light traffic. For these facilities, gravel thicknesses may be 0.5 m or even less. Bearing capacity will be low and heave and thaw settlement are high, so the roads are very expensive to maintain. It proves economical in the zone of sporadic permafrost where there are only a few isolated icy masses that cannot readily be predicted, and also in hilly parts of the continuous and discontinuous permafrost zones where large fills have to be emplaced and suitable frost-stable material is not available. An example of the latter is the section of the Dempster Highway through the Richardson Mountains east of the Rat Pass. The long term effects of this type of construction on the permafrost table can be immense (Fig. 5.5).

It is possible to place compacted frost-susceptible soils such as silt or clay in a waterproof membrane to maintain a given density and strength in the soil (Joseph & Webster, 1971). This is called the "membrane-enveloped soil layer (MESL) technique". It uses local materials, compacted and sealed, as a base for the road and appears to work satisfactorily (Schaefer, 1973) if the seal is not broken by stones or snowploughs (N. Smith & Paszint, 1975). Should the top of the membrane rupture, water penetrates and cannot escape, so that severe heaving follows. This probably explains why it is not generally used. The consequences of deep thermal cracking which occurs across roadways in cold regions is also undetermined, but will probably cause breaks in such a membrane system.

Winter construction is ideal as long as there is minimal snow cover. Both the active layer and fill are frozen, so maximum strength is achieved as soon as the pad is compacted, making it possible to open the road or railway several years earlier than with summer construction. Until the surface under the pad has frozen, the bearing strength of the road is low and the road cannot be used. With summer construction, strength will gradually increase as the ground and pad get colder.

Should it be decided to pave the road, this should not be done until adequate bearing strength has been attained, and unless it has been determined that the fill is thick enough to prevent thawing after surfacing. The use of a light coloured surface aids in reducing this problem (Fulwider & Aitken, 1962; Wechsler & Glaser, 1966; Kritz & Wechsler, 1967; Berg & Quinn, 1966; Berg & Esch, 1983). However, most surfaces tend to break up with movement of the underlying bed. The use of a thin spray-and-bind coating that is flexible and will move with the underlying surface has proved very successful on the Alaska Highway since 1982. Failure of the surfacing only occurs where heaving and settlement are excessive. The major advantages of successful paving are that it produces a reliable all-weather surface which requires minimal maintenance, and tyre problems on vehicles using the road are virtually eliminated. The main disadvantages are the need for a fairly stable, good quality pad underneath it and the initial cost.

Actual Construction

In areas of sporadic or discontinuous permafrost, construction may be preceded by clearing and deliberate thawing of the permafrost. In some cases, the new right-of-way is cleared several years ahead of construction and the surface vegetation destroyed to cause rapid thawing. This is used along the

Alaska Highway on the plains west of Fort Nelson. In all other cases, clearing of timber and other vegetation is carried out as construction proceeds to minimize the amount of thawing. Any burning must be carried out where the embankment is to be built to reduce the tendency to develop a thaw pond. Esch (1982, 1984) has shown that addition of an asphalt-covered pad overlain by a clear polyethylene sheet produces more rapid thawing.

Construction is carried out in such a way as to minimize the amount of fill needed and yet avoid seasonal thawing of more than the pavement and pavement base. If the underlying ground remains frozen, this minimizes settlement of the ground and fill and so keeps down maintenance costs.

The primary cause of seasonal heaving and thaw settlement is the accumulation and retention of water in the seasonally thawed zone. By using a very free-draining base to the fill, capillary action is reduced to the minimum so that bearing pressures that can reach 1.3 million Nm^2 will be kept at values where the weight of the overlying fill will not exceed them. Clean crushed rock or gravel is ideal. Even a small quantity of silt or clay can cause undesirable performances. Depth of fill is particularly critical where cuts must be made in ice-rich, silty soils, and construction should only be carried out in winter. On coarse subsurface materials, stability may be achieved without using large volumes of overlay. Where there is less than 0.6 m of organic soil, this may be removed and immediately replaced with granular fill. If this is not done in thawed muskeg, the soil tends to compress and flow laterally beneath the weight of the fill. Overlay construction on the undisturbed active layer is carried out in summer.

Roads built over bedrock can cause problems due to the tendency for silt and clay to have accumulated above the bedrock surface, possibly maintaining a perched water table. In Canada, it is usually considered desirable to remove the soil, wash out the silt and clay and replace it by granular fill. This is not done in Alaska.

Muskegs or peaty swamps create special problems. Fortunately the thickness of peaty beds tends to decrease at higher latitudes (MacFarlane, 1969). It is essential to determine the thickness of the organic layer, the type of underlying soil, and the depth of seasonal and perennial frost in open areas. More than 25-46 cm of frozen ground will usually support construction equipment. Removal of the snow cover will aid in increased frost penetration. Ideally, any ice wedges present should be identified in advance, although this is usually impractical.

The trees and shrubs are removed with as little disturbance of the organic soils as possible. If the surface is weak, then logs or trees must be laid as a corduroy matting to spread the weight. Next, a brush or wire mesh may be laid on the logs to avoid the fill settling through the layer. However, it must be remembered that rotting of brush or corrosion of mesh may cause problems later. In general, the organic soils overlie either bedrock or gravel, or else silts with a high ice content. Thin, organic soils are often completely removed prior to replacement with fill. In the case of underlying gravels or rock, the fill can be placed directly on the organic soil, which will compact and not tend to flow or creep laterally.

In almost all construction on permafrost, end-dumping of fill is employed to advance the road base along the right-of-way. Over ice-rich silts, the ground must be kept frozen as long as possible, and at least 2 m of fill needs to be placed on its surface in the first place. Without this, the ground thaws and the fill will sink into the resulting thermokarst pond. Over permafrost in peat, Alaskan construction practice is to make every effort to avoid disturbing the organic mat or displacing the peat. Frozen peat is considered an ideal foundation soil for embankments and it has even been placed beneath a roadway to preserve the underlying permafrost (McHattie & Esch, 1983).

Drainage

Drainage of the land alongside the road must be adequate to remove the water coming off the road and maintain the ground-water level well below the road base, even at the maximum distance from the drains. Ditches must have sufficient slope to remove all the surface water or else ponding may result in melting of ground-ice and the potential for the development of thermokarst lakes. Lining of the ditches may be necessary to avoid erosion and these ditches should be located as far away as possible from the shoulders of the roadway. Minor ponding may not be detrimental (Esch, 1983).

Wherever possible, roads should be built parallel to the direction of groundwater movement to aid in draining the land. Culverts are normally emplaced to

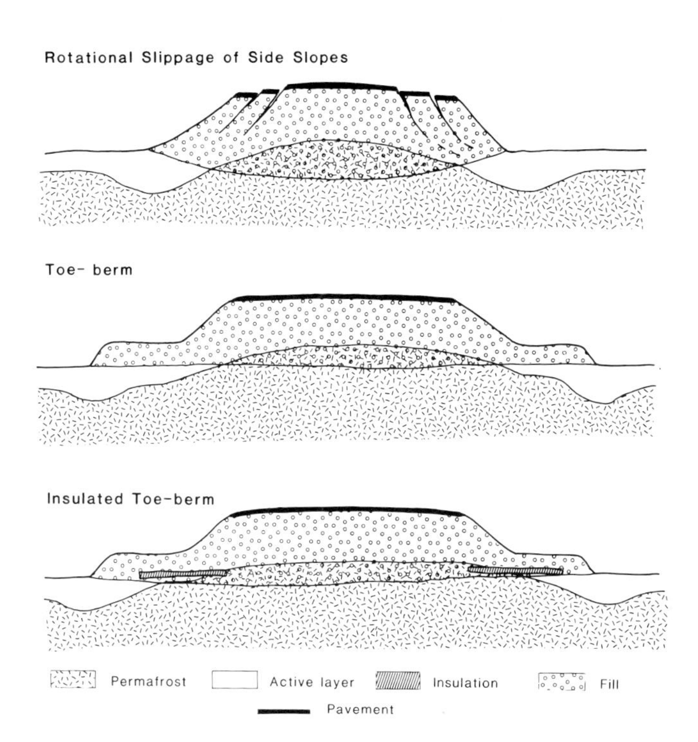

Figure 5.6: Comparison of permafrost tables in metastable and stable permafrost (<2°C) below insulated toe-berms, toe-berms, and where rotational slippage of side slopes of embankments has taken place (modified from Esch, 1983). See Plate XI.

deal with cross-drainage, and they should be planned to be adequate to handle the highest anticipated runoff during snow-melt. It must also be remembered that the road may act as a dyke to groundwater-flow because of the elevated permafrost table in the road base where a high fill is used. This extra water must also be accommodated in the culverts. Culverts should be placed at all low points on the landscape and extreme care should be taken at stream crossings to position the culvert so as not to alter the stream flow regime. Some channel stabilization may be necessary upstream in the form of dykes to avoid problems with channel migration.

Too little fill results in a lowered permafrost table under the road. This results in a loss in bearing strength which must be allowed for in the design.

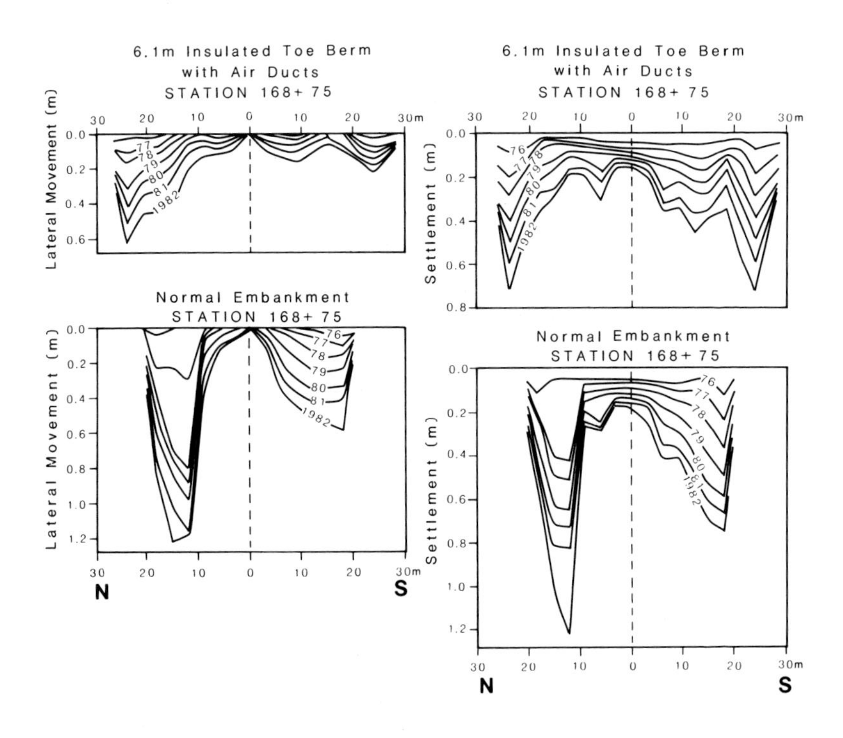

Figure 5.7: Lateral and Vertical Movements of Normal and Bermed, Insulated and Ducted Embankments, from Reference Points at Various Distances Left and Right of Roadway Centreline (after Esch, 1983, Fig. 8).

Metal culverts are the most successful in muskeg due to their ready adjustment to and tolerance for settlement, since concrete culverts tend to break up under those conditions. Aluminium or wooden culverts do not corrode in the acid environment of muskegs like steel or concrete (MacFarlane, 1969). The openings in culverts carrying no water in winter (MacFarlane, 1969) should be closed to prevent plugging with snow where wind-drifting is a problem. In spring, they must be reopened to permit drainage of the snow melt. Livingston and Johnson (1978) showed that insulation of the culvert plus use of boulders to reduce cold-air circulation in the openings can be effective in controlling icing of the culverts.

Embankment Problems

Even where drainage is adequate and the road base becomes frozen above the original ground surface beneath the centre of the embankment, problems may occur. These are generally due to thawing along the embankment edges and various precautions have been suggested. G. H. Johnston (1981) recommends 3:1 slopes where road embankments are less than 1.8 m high, and 2:1 where heights are greater. This has the value of conserving fill but it also increases the risk of trouble on the side slopes. On railway embankments, 3:1 or more gentle slopes are used to obtain greater stability.

A common problem is the development of rotational slips along the margins of the pavement (Fig. 5.6), the evidence for this consisting of longitudinal cracks developing parallel to the road margins (Plate XI). To overcome this tendency, it has become common practice to put in toe-berms, often with their own layer of insulation to try to reduce thawing of the ground along the base of the embankment. The latter tends to take place due to increased snow depth in winter and runoff water in summer.

Esch (1983) describes the results of a study of the performance of two partially insulated experimental embankments protected by insulated toe-berms and air cooling ducts over a period of seven years on the Alder Creek-Parks Highway and Bonanza Creek near Fairbanks. Although the permafrost had penetrated the main body of the fill, mean annual surface temperatures reached 3°C on the south-facing slope and 1.5°C on the north-facing slope. Where insulation was placed at a depth of 1.2 m in the fill, there was least settlement, but the same insulation laid at 3.2 m gave appreciably poorer results (Fig. 5.8). Progressive thaw-related deformations, accompanied by creep movements of frozen soils have resulted in cracking of slopes and shoulder areas, varying amounts of settlement of the pavement (up to 2 m) and spreading of the embankment by up to 1.8 m (Fig. 5.7). No combination of air ducts, insulation, and toe-berms was adequate to totally prevent the movements. Thus it is necessary to allow for rather large progressive movements in the design of embankments.

Icings

These represent a major problem along many roads and railways. In the former case, vehicles with chains can negotiate short stretches of ice, but

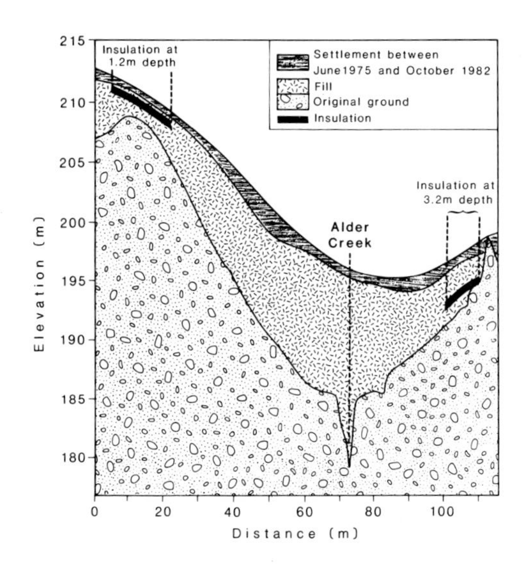

Figure 5.8: Combined Effects of Settlement and Creep in Seven Years Along an Experimental Embankment Protected by Toe-berms, Insulation and Ducts on the Alder Creek - Parks Highway at Alder Creek, Near Fairbanks, Alaska.

icings must be completely prevented or thawed on railway tracks if trains are not to be derailed.

Icings develop where spring, surface or seepage water flows over the surface. When temperatures are below freezing, this water tends to freeze on the cold ground surface, coating it with ice. Additional layers form as the water flows over the ice surface. A build-up of 2.4 m of ice is not uncommon (van Everdingen & Allen, 1983), while along rivers in Siberia, they may extend for 100 km. Along the Canning and Sagavanirktok rivers (Harden et al., 1977) in northeastern Alaska, they are between 40 and 50 km long. The largest ones are associated with braided streams.

Pre-existing icings can be recognized on aerial photographs taken in winter (Harden et al., 1977; Åkerman, 1982), but long-term studies show that icings do not form in exactly the same place, nor are they the same size each year (van Everdingen, 1982; Slaughter, 1982). The construction of a highway across drainage and seepage zones intensifies the icing problem due to the road embankment acting as a dyke, holding back the water. Culverts are apt to freeze the surrounding soil as water freezes in the roadbed. Bridges cast shadows which act as loci for icings in winter, hence the need to place all bridges well above the anticipated water level.

Carey (1973); Carey et al. (1975); Johnson and Esch (1977); Ye Bayou and Yang Hairong (1983); van Everdingen (1982) and Thomson (1963) discuss various means of combatting icings. Thomson divides the methods into active methods and passive methods, the active methods being those avoiding the development of icing problems, while passive methods are those helping to minimize the effects of icings.

The active methods start with examining the icing pattern in a region and locating the roads where no active icings can be detected. If this is impractical, the locus of the icings for a succession of winters should be identified so as to determine the probable cause, the pattern and degree of variability of the natural icings. Based on this knowledge, oversize culverts can be placed in the key locations to carry the water flow related to the icing. These should be located on a steep gradient as deep in the icing as possible. Small steam pipes can be placed in the culverts with their ends projecting upwards, well clear of the maximum anticipated thickness of the ice. The top ends of these pipes are capped to keep out the snow until they are needed. Large relief culverts should be placed above the main culverts through the fill so that these can carry the flow of water when the main culverts are filled with ice. The grade of the road may be raised to lift it above the anticipated maximum level of the icing.

If the volume of the icing is known and is fairly constant, it is possible to place a dyke upslope, parallel to the road, and often with an excavation that is of a suitable size to contain the entire icing. This has proved much more successful than freezing belts, which are zones upslope cleared of the insulating vegetation mat and of snow. The cold penetrates the ground and freezes the water at that location instead of the road. In spring, the vegetation mat must be replaced to try to prevent thawing of the ground. A berm or dyke is usually necessary to achieve this effect.

It is also possible to place insulated rock-filled drains in the ground with a sandy covering. These can collect and transmit the seepage water that would normally form an icing. van Everdingen

(1982) describes an example from 3 km south of Haines Junction, Yukon Territory, which is apparently performing adequately.

Finally, icings at bridges and culverts can be reduced or even eliminated by modifying the channel so that the water travels in a narrow, steep-sided channel with rapid flow. Braided channels are the most susceptible to icing.

The passive methods include thawing an opening through the culverts, using portable steam generators or hot-water boilers mounted on trucks. These are a regular sight on many highways in permafrost regions. Sometimes hessian cloth is hung across the surface of water seepage in a vertical sheet. Ice builds up on it and behind it, instead of against and over the road. Graders can be used to rip off the ice as it forms on the road surface. This is a common practice along the Alaska Highway but has to be carried out almost daily, which is very expensive. On roads where the ice build-up is excessive and regular grading is impossible, holes can be steamed in the ice and dynamite used to break up the mass. Then the broken ice must be removed immediately and the surface smoothed with a grader. Finally, where there are small icing problems in culverts, fire pots made of 200 litre drums and burning continuously, may keep the mouths of culverts ice-free. They are a common sight in central Alaska and the Yukon Territory but are not normally used in Canada since they need regular attention and consume from 40 to 100 litres of fuel per day. These firepots or "moose-warmers" do prevent icing up of the culvert under the road since they heat the water entering the culvert.

Alternative methods of thawing the ice in culverts include using electrical heat tape, pumping hot water from a boiler mounted in a truck through the steam pipes and since 1981, the use of a solar-powered thawing device. The latter is by far the cheapest method and was developed at Fairbanks by the State of Alaska Department of Transportation and Public Facilities (Sweet, 1982). It uses a flat place collector, a twelve volt circulating pump and radiator hose. It successfully kept water flowing through a culvert from the removal of its covers in March until the snow had melted in May. Ethylene glycol is used in the system.

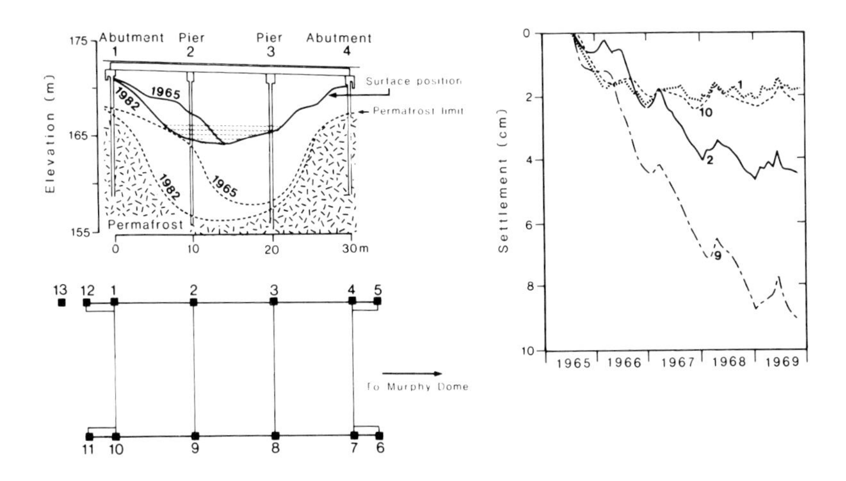

Figure 5.9: Cross-section and Displacement of Selected Piles with Time, Goldstream Creek Bridge (after Crory, 1985). See Plate XII.

Bridge Footings

As already noted, the usual method of crossing large streams and rivers in permafrost areas is by using ferries. Crory (1985) provides an example of the problems encountered with footings for bridges. Underneath the water course there is usually a talik. The adjacent permafrost is usually very ice-rich. When the stream moves, the position of the talik changes so that unless the stream course has been stabilised, this will result in thaw subsidence in the ice-rich alluvium. Where the stream moves away, the deep talik tends to show aggrading permafrost which is accompanied by heaving as the ice develops in the former talik. The larger the river, the more difficult it will be to stabilise its course.

Plate XII shows the Goldstream Creek bridge, built in 1965, near Fairbanks, Alaska, viewed from the west. Crory (1985) has shown that in the 18 years to 1983, the stream channel had changed position and that this was accompanied by thawing of the ice-rich permafrost (Fig. 5.9). As a result, shims had to be placed under the bridge on top of the piers and these shims adjusted regularly as the piers subsided. Fortunately the bridge does not carry heavy loads, but clearly piers 2 and 9 are ceasing to be firmly anchored in permafrost.

Stability of Cut Sections

One of the major problems in permafrost regions is how to deal with cut sections in ice-rich permafrost (Smith & Berg, 1973; Pufahl et al., 1974; Jackman, 1974; Berg & Smith, 1976; Murfitt et al., 1976; Wang et al., 1977; Shu Daode & Huang Xiaoming, 1983). In competent bedrock, the only problems are normally with perched water tables above the permafrost table, and with blocks of rock falling from the 4:1 face. These can normally be coped with since there should be a 2.5 m wide drainage ditch at the base of the slope. Free-draining granular soils also behave satisfactorily.

The normal practice in ice-rich cohesive sediments in the past (Berg & Smith, 1976) has been to cut slopes nearly vertically and then to allow the slope to thaw and slump. The overlying vegetation mat drops down and forms an overhang and prevents the underlying ice-rich material from flowing out during thaw. Trees should be cleared by hand from the zone undergoing thawing and slumping. Generally, the slope stabilises when it has changed to about 1:1.5, although on occasion even this slope is too unstable. Good examples can at present be seen along parts of the roads in central Alaska along the Alyeska Pipeline. Although problems with gullying can occur, the slope will revegetate naturally in as little as three thaw seasons.

Where problems are anticipated, various alternative techniques can be employed (Fig. 5.10) to reduce the maintenance costs and provide a more stable slope (Pufahl, 1976). These include cutting the back slope to a steep face and piling gravel against it and above it on a 35° slope (A). Insulation can be added near the cliff top to reduce thawing. However, this is expensive and uses tremendous quantities of gravel. If the cliff is stepped and the gravel applied as a veneer with a layer of insulation underneath, a similar effect can be produced (B). Steeper slopes are possible if the gravel is mixed with rock and retained in wire gabions, although a toe-berm is necessary to hold the gabions in place (C). If gravel is plentiful and the cost of insulation is not prohibitive, then the slope can be reduced to 20° as in (D). The main drawbacks of all these methods are the initial high costs and the difficulty in revegetating a sloping gravel surface. It remains to be seen whether these high cost methods will be used to any great extent.

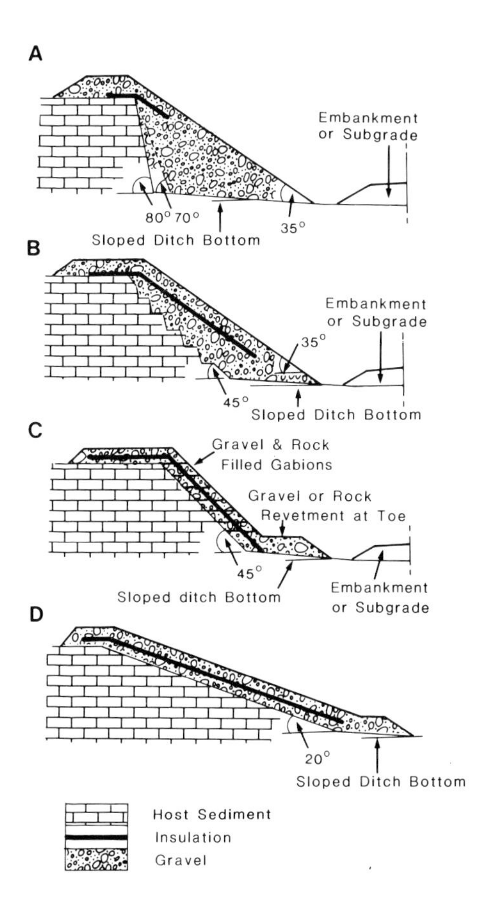

Figure 5.10: Possible Methods of Stabilising Cut-Slopes in Ice-Rich Cohesive Sediments (after Pufahl, 1976).

SPECIAL CONDITIONS FOR RAILWAYS

It will be seen from Figure 5.1 that in North America, railways are not normally built on continuous permafrost. Railways are only put in place when there is a demand, usually based on industrial need. This need implies that the railway must be built quickly and put in service in minimum time. This means that staged construction is impossible, even though it is used on roads, e.g., the Dempster Highway and on the haul road of the Alyeska pipeline.

These were only opened to public traffic after the road had acquired adequate strength to take the pounding from the loads.

The cross-section of the railway embankments has to be modified to have a wider base, gentler side-slopes (1:3 maximum) and toe-berms, to reduce the tendency for creep and settlement soon after construction. Fortunately, it is relatively easy to lift the rails and place more fill and ballast to correct for differential settlement, and this represents a major portion of the maintenance costs. Individual parts of the track can be shimmed.

General methods of construction are similar to those for roads but icings must be avoided if possible. Experience with construction of the existing railways is described in Charles (1959, 1965), A. V. Johnston (1964), Pryer (1966) and Woods et al. (1959). They were mainly built between 1910 and 1960 and the first ones used hand labour to a great extent. Modern methods such as using thermal piles are now being used to stabilize sinkholes to stop settlement (Hayley et al., 1983) rather than concentrating on filling in the holes at regular intervals.

Similarly in China, the railways are mainly found in the sporadic and discontinuous permafrost zones. Considerable care is being taken to determine how rail construction can be improved and the Ministry of Railways is one of the two main organizations spearheading permafrost research (e.g. Shu Daode & Huang Xiaoming, 1983, Nei Fengming, 1983; Li et al., 1983; and Huang Xiaoming, 1983).

In Russia, the railways generally stayed outside the zone of continuous permafrost until about 1950. Then, with tension between China and Russia, it was decided to build the Baikal-Amur-Manchuria line further from their common border. The "B.A.M." railway was completed in 1984 and it remains the main rail line on continuous permafrost. Unfortunately, details of the construction methods are not available but the time taken to complete it suggests that staged construction was necessary. This would also occur when opening up Public Lands in North America.

In the zone of sporadic permafrost, summer construction is possible and an attempt can be made to pre-thaw the permafrost islands to reduce long-term maintenance costs. As more permafrost is encountered, it becomes necessary to try to preserve it, and winter construction is essential for at least the road base. This is placed over the frozen ground with a minimum of surface disturbance. Typical minimum thicknesses for the pioneer fill would be 1.8 m in the forest and 2.7 m in the tundra. Additional fill is emplaced during the following summer.

In Figure 5.1, it will be seen that almost all the railways in North America cross permafrost in lowlands. This reduces the problems with cut-and-fill but still leaves drainage and river crossings as major problems. Bridges are much longer for railways due to the need for maintaining a nearly horizontal grade for the track.

The only recent attempt at putting a railroad through mountains over permafrost was by British Columbia Railways Co. Ltd. from the Cassiar mine south along the Cassiar highway. After spending 8 million dollars and encountering severe problems south of Dease Lake, the project was abandoned. The reason appears to have been that the design did not allow for the permafrost.

A more successful railway runs from Anchorage to Fairbanks over flatter terrain. The main problems of maintenance on that railway were demonstrated during the Fourth International Permafrost Conference (Fuglestad, 1985). Construction commenced in 1904 at a time when frost-susceptibility of materials had not been studied. Use of any convenient material as fill resulted in considerable localized seasonal heaving in many places that must be dealt with. This is partly done by using special shoes for the rails that can be adjusted for up to 20 cm of vertical movement without normal shimming. When shimming is carried out, spikes must be removed, the holes in the sleepers (called "ties" in N. America) are filled and then the spikes replaced. This can reduce the effective life of a sleeper to six to eight years from the average life of 35 years.

Chapter Six

AIRFIELDS

The development of aeroplanes had a tremendous impact on opening up permafrost regions. First were the float planes that could land on lakes, rivers and on the open water of the sea during the short summer. Then, by fitting skis, the flatter ice surfaces could be used in winter. However, in order to bring in large quantities of supplies and equipment, either temporary winter or permanent all-weather runways had to be built. In both North America (Fig. 6.1) and the Soviet Union (Fig. 6.2), these have become a major factor in the transportation network for the vast regions not served by all-weather roads or railways.

Thus, in Canada, the first phase of summer flights by bush pilots lasted from 1924 to 1938. Aeroplanes were used for oil exploration in the Mackenzie valley around Norman Wells. By 1938, some settlements had built small community airstrips to handle regular air mail and small-scale passenger services.

World War II brought extensive airstrip construction along both the Alaska Highway and the Mackenzie valley. Existing airstrips were upgraded and the Eastern Arctic started to open up, with airstrips being built at Churchill, Coral Harbour, Fort Chimo and Frobisher Bay. Five "Joint Arctic Weather Stations" were established at Alert, Eureka, Isachsen, Mould Bay and Resolute Bay after the war and were supplied by annual military airlifts.

Soon afterwards, the Distant Early Warning Line radar stations were added, each with its own airfield. Currently, the resource industries have increased the use of air transport in the region, and drilling rigs are moved to and from isolated sites by aircraft using specifically constructed runways. Airstrips are also located at strategic

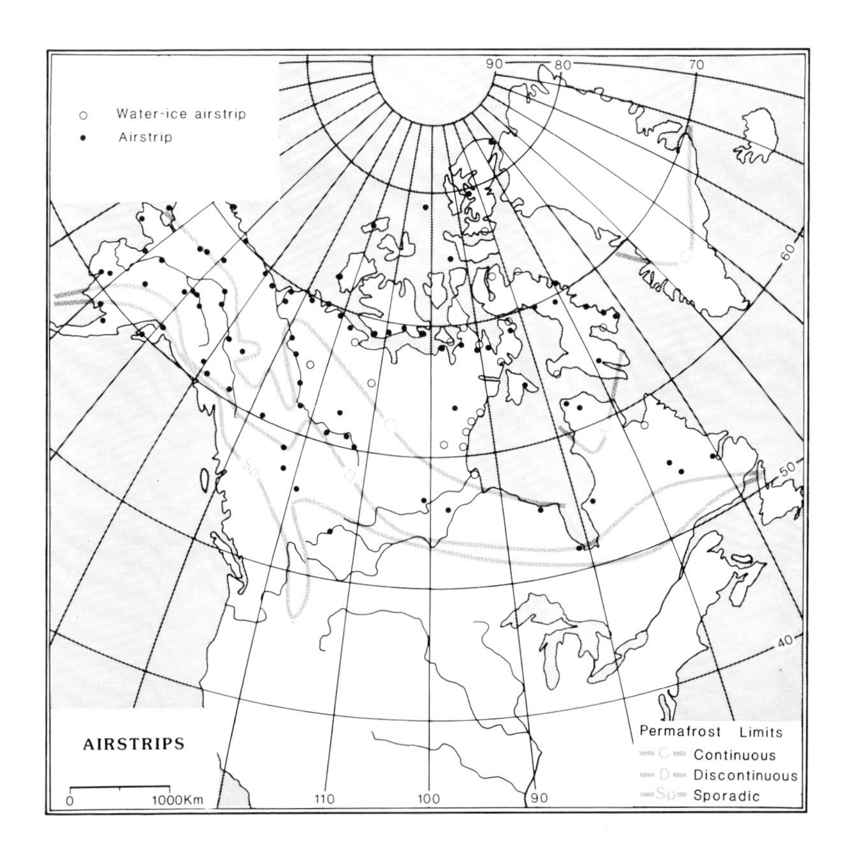

Figure 6.1: Main Airstrips in the Permafrost Regions of Northern North America.

points along pipelines to aid in servicing them, e.g., along the Alyeska Pipeline in northern Alaska. Interchangeable wheels, skis and floats on small aircraft help in this, while heavy freight can now be brought in quickly in the larger aircraft.

PERMANENT AIRFIELDS

These are the centres for economic development in their environs. Three classes of airports are recognized in Canada, the highest grade being Arctic A airports consisting of a paved runway with a minimum length of 1830 m and a minimum width of 45 m. Arctic B and C runways have a gravel runway

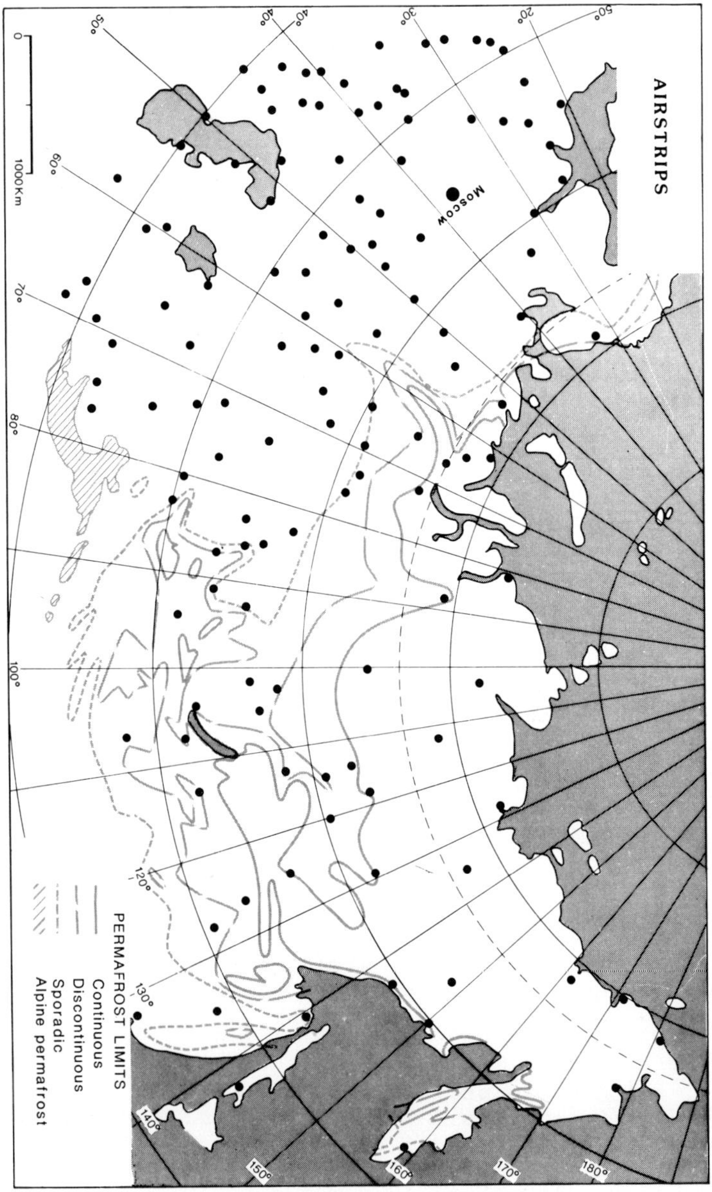

Figure 6.2: Main Airstrips in the Permafrost Areas of Eurasia (data from USSR, 1977, p. 19; USSR, 1980, Fig. 168).

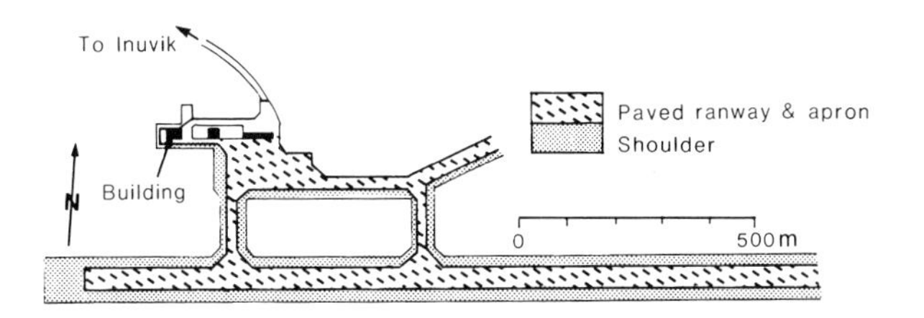

Figure 6.3: Layout of Inuvik Airport, Northwest Territories (after G. H. Johnston, 1982, p. 578). Reproduced with the permission of the National Research Council of Canada.

with minimum lengths of 1525 m and 916 m and minimum widths of 45 m and 30 m respectively. The class of airport required depends on its potential use.

A typical airport layout for a main regional centre would be that of Inuvik airport (Fig. 6.3). This is an Arctic A airport and its construction has been described by Johnston (1982). The basic method of construction is that of a pad, often with insulation to reduce the thickness of fill required. It is essential to ensure that the permafrost table rises into the base of the fill so that the former ground surface remains permanently frozen. Allowance must be made for the thermal effects of paving if this is to be the final surfacing.

In airfields, approach angles and prevailing wind directions are very important (Harwood, 1966). In addition, proximity to towers and radio-communication arrays must also be considered. After these considerations are accommodated, an adequate source of good fill must be located. If this is readily available, the soil conditions are of

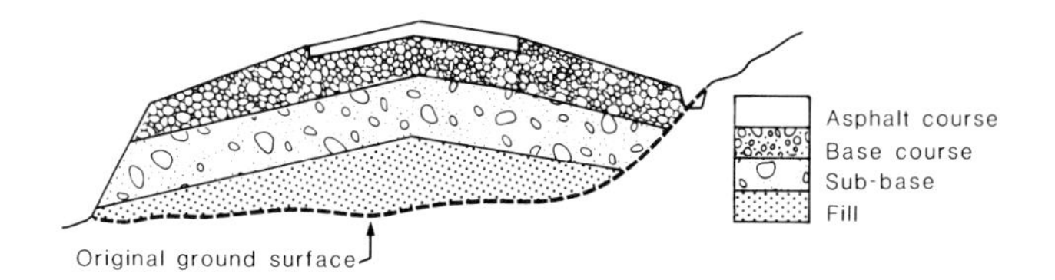

Figure 6.4: Typical Cross-Section of an Arctic A Airstrip (G. H. Johnston, 1981, p. 389), vertical scale exaggerated. Reproduced with the permission of the National Research Council of Canada.

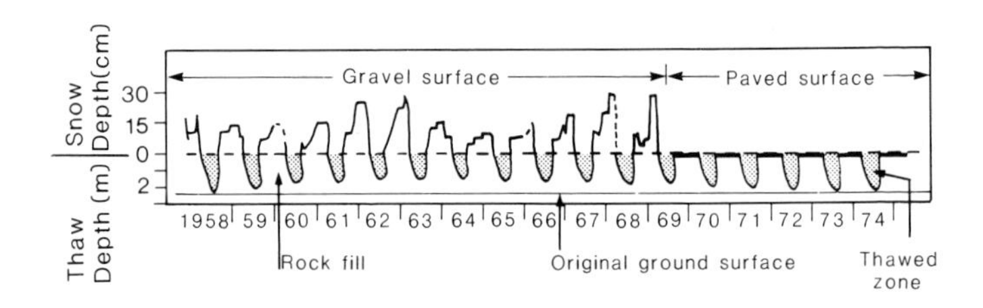

Figure 6.5: Snow Cover and Thaw Depth Below the Runway Centre-Line at the West End of Inuvik Airstrip (G. H. Johnston, 1982, p. 582). Reproduced with the permission of the National Research Council of Canada.

relatively little importance, although well-drained sites on coarse, frost-stable material are ideal. In spite of this, given enough good quality fill such as beach gravel, airstrips have been built over frozen ponds and lakes up to 1.2 m deep in winter, without the ice thawing subsequently (Harwood, 1966). Figure 6.4 shows a typical cross-section of an Arctic A airstrip.

In the case of Inuvik, crushed rock was brought in from a quarry 0.8 km to the west and was used as fill. The land was first cleared by hand, and then the fill in the subgrade was laid in the winter of 1958 over the frozen ground. The base was added during the next summer. Insulation was used to reduce the quantity of fill required (Brown, 1970a). Thermistors were installed to check on the thickness of the active layer under the airstrip and it will be seen from Figure 6.5 that the underlying ground surface has remained frozen since 1958. The base course was sprayed with liquid asphalt to keep down the dust.

It usually takes about five years to ensure that the airstrip is adequately frozen so that paving with asphaltic cement can be carried out. Figure 6.5 shows that there was an increase in depth of the active layer beneath the runway at Inuvik of up to 40% after paving.

The actual thickness of asphaltic concrete and base depends on the tyre pressure of the aircraft (Table 6.1). Since loads are getting heavier, there has been a need to upgrade the more important runways with the increase in use. Side-slopes are normally about 1:10.

Table 6.1: Typical Layer Thicknesses Used in Canada for Asphalt, Base and Sub-Base Courses (after G. H. Johnston, 1981, p. 391). Reproduced with the permission of the National Research Council of Canada.

	Minimum Layer Thicknesses (mm)			
	Design Aircraft Tyre Pressure (kPa)			
Layer	<415	<690	690 to 1030	1035 to 1370
Asphaltic concrete	50	65	90	100
Base	150	230	230	305
Sub-base	- As necessary to provide adequate bearing strength in flexible pavement.			

During construction, all the precautions used for roads and railways must also be employed. In general, the higher the embankment, the greater the cooling and the fewer the drainage problems. The use of crushed rock in the fill minimises upward water movement by capillarity.

At the main airstrip, terminal buildings will be necessary and these will have to be constructed using the techniques described in Chapter 4. Haul roads for fill will normally be made of thick pads of fill.

The consequences of failure to take care of the environment during the construction of an Arctic airstrip have been described by French (1975b). At Sachs Harbour, Northwest Territories, the airstrip was built on a ridge with a saddle in it. In order to try to fill in the low spot, bulldozers were used to push the nearby thawed active layer onto the low spot during three successive summers. As a result, progressive subsidence has occurred with the development of spectacular thermokarst mounds due to melting of underlying ice wedges.

TEMPORARY OR LIMITED-USE AIRSTRIPS

When an airstrip is needed for temporary use, e.g., to enable an oil well to be drilled at an inaccessible site, access is gained originally by a winter road (ice-capped or ice-road). For an oil rig, an Arctic B airstrip is necessary and bringing in the drilling equipment will involve about 150 trips by a C130 Hercules aircraft. In addition, men and supplies will be brought in and the rig must be removed on completion of drilling.

The airstrip must be sited at a suitable distance from the proposed drilling site, and availability of fill is critical. Construction is carried out in winter by laying a thin cover of suitable fill, followed by 5-7.5 cm of multilayer insulation. It is covered by an impermeable membrane and then by more fill. This fill is laid and compacted in the frozen state. Actual thickness of insulation and fill required is calculated from the air temperatures, the soil temperatures and the precipitation (see Johnston, 1982). Sometimes aluminium military matting, painted white, is laid on top of the fill. Then the airstrip is ready for use.

After drilling is completed and the drilling rig is removed, the airstrip is normally sown with short grasses to stabilize the surface and it can then be used again whenever necessary. Problems sometimes develop along the shoulders of the airstrip as they do along roads (Chapter 5) and the same precautions must be taken to avoid trouble.

For airstrips used during only a single season, e.g., along oil pipelines, an ice strip is commonly constructed. If the ice is thick enough, it will prevent environmental damage and it will be difficult to see that it was ever there. These were extensively used during the construction of the Alyeska pipeline.

Chapter Seven

OIL AND GAS INDUSTRY

The vast permafrost region includes areas with considerable potential oil and gas reserves as well as proven deposits. As conventional oil and gas reserves in more accessible places are used up, more and more attention will be paid to finding and using deposits from permafrost regions.

EXPLORATORY AND SEISMIC SURVEYS ON LAND

Oil has been known in the north since Sir Alexander Mackenzie reported oil seeps along the Mackenzie river near what is now Norman Wells in 1789. Leffingwell (1919) reported similar oil seepages near Cape Simpson, Alaska. The early exploratory work was carried out by dog-team and boat, and the first oil well in the region around Norman Wells was drilled in 1924. From then on, the oil reserves there were found and developed by modifying methods used elsewhere to cope with the special problems caused by permafrost. Aeroplanes opened up the previously poorly accessible North. However, the oil potential of the North Slope of Alaska was not seriously studied until after World War II.

In all cases, the bulk of geophysical work on land must be carried out using winter trails and roads (see Chapter 5). Properly constructed and maintained, the only evidence in summer of where they are located are the long, straight clearings running across the landscape, although these may still cause thermokarst to develop near the southern limit of permafrost. The main environmental problems that arise from seismic surveys are due to the need to place the geophones on a firm surface. If the blade of a bulldozer is used to remove the loose snow, it is liable to slice off the tops of hummocks and mounds (Adam, 1978). This also can result in the development of thermokarst. Since 1974, there have been fairly precise regulations regarding the use of seismic trails which have largely overcome the potential for thermokarst problems in North America.

When a suitable geological structure is found, the winter roads will be used to take in small drilling rigs in order to conduct geotechnical site investigations to determine the near-surface stratigraphy and ice content, so that the best sites can be selected for deeper drilling, for obtaining frost-stable fill material, and for locating the associated buildings, water supply and sump.

DRILLING RIGS

The drilling rigs used in oil and gas exploration are very large. When disassembled, they require about 50 large trucks to move a single rig. In summer, gravel roads need regrading after every four loaded trucks have passed. Winter roads are preferred because they are cheaper and cause less environmental damage. The alternative is to fly the drill rig into the chosen site and to airlift in men and supplies as necessary. This requires an adequate airfield to handle the traffic.

The drill must be placed on a suitable pad which will not settle or move during the drilling operation. The drilling process requires the use of a circulating fluid to lubricate the bit and this fluid must be stored in a suitable reservoir. The latter is commonly excavated or blasted in the ground as close as practical to the rig. The volume must be adequate to meet all the needs of the drilling operation, even if it is found necessary to drill deeper than originally planned. Losses from seepage must be allowed for. The drilling fluid is normally water to which various additives such as barium sulphate (barytes) have been added. These additives must provide sufficient weight to prevent water, oil or gas under pressure from blowing out of the well, and they must seal the well bore against fluid losses. Additives such as calcium chloride may also be necessary to counteract the tendency of the fluid to freeze in the upper layers of the ground, and a suitable source of water for the fluid must be found.

Obviously winter is the preferred time for drilling, since it minimises the risk of the ground thawing beneath the rig but it makes it more difficult to keep the drilling fluid from freezing.

Upon completion of drilling, the drill must be disassembled and removed and the site restored to its original state. This involves filling in the sump in which the drilling fluid was stored and minimising the environmental effects of the fluid and its constituents until they are completely frozen and have become part of the permafrost.

Offshore, a suitable drilling platform is necessary which may consist of thickened ice or a man-made island. The rig must be transported to the site. Water is easier to find but storage must be in tanks on the island. The drilling fluid is usually discharged into the sea when it has been used. After completion, the well must be either plugged and abandoned (no further use) or plugged and suspended. Suspended wells are plugged in such a way that the well can be re-entered, the plugs drilled out, and the well completed for production. These suspended wells are referred to as "keepers". Casing for production wells through permafrost may experience much more long-term thaw subsidence and related strain than the exploration wells. This means that if an exploration well is to become a keeper, this must be planned in advance, in order that the casing design is adequate for production at a later date. Finally, if drilling through the ice, the drilling must be completed quickly before the ice platform moves away.

DRILLING OIL AND GAS WELLS ON LAND

Over 200 land-based wells were drilled in Arctic Canada between 1959 and 1979. Anon. (1982, pp. 4.50-4.65) discusses the methods used in drilling through permafrost, together with the main problems commonly encountered. Figure 7.1 shows schematically the special precautions taken for producing wells on land where permafrost is present (Anon., 1982, Fig. 4.4-6).

The first problem is the presence of thaw-sensitive permafrost, necessitating avoidance of thawing of the ground. Adequate pads must be placed under any roads or airstrips, as well as under the rig itself to avoid thawing of the ground. The sump must be constructed so as not to thaw the ground under the rig. An impermeable membrane and matting may be used. Insulation may be applied around the casing of the well, or an extra-casing with mechanical insulation may be installed to create a "thermos" effect. The temperature of the drilling mud is kept near freezing and the well is drilled quickly

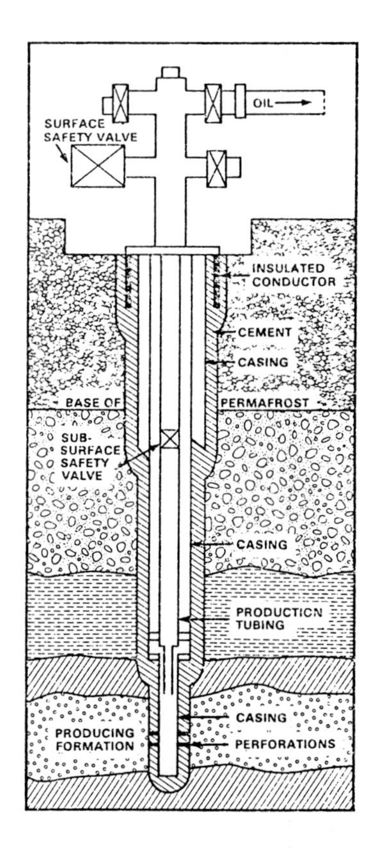

Figure 7.1: Special Precautions Taken for Wells in Permafrost (Anon., 1982, Fig. 4.4-6).

through the permafrost zone. Special cements are used that set at low temperatures for cementing the upper strings of casing to the well. The casing must remain intact even if thawing of the permafrost takes place. Should the well be turned into a producing well, it will require the precautions shown in Figure 7.1.

Production of warm oil and gas will cause thawing of permafrost around the casings. The soil will deform (thaw subsidence), inducing stresses in the casings and causing some ground surface settlement. The designer must determine the properties of the permafrost (before and after thaw), the radius of the thaw zone for the design life of the wells, and the stress/strain properties of the well casing.

The designer must then calculate the stresses/ strains that the casings would endure if the thawing was allowed to occur unimpeded. If this is tolerable, the design is satisfactory. If the casing strains would be intolerable, then measures such as insulated casing, insulated tubing and possibly refrigeration must be considered.

In addition, precautions must be taken in case gas hydrates are encountered in the upper 1500 m. The gas content of the drilling fluid is monitored and if notable gasification occurs, gas hydrates are probably present in the walls of the well and remedial action must be taken (Goodman & Franklin, 1982). Slower drilling, chilled mud and greater mud weights are usually adequate to control the problem, the idea being to change the hydrate equilibrium conditions on the sides of the hole so that further decomposition is prevented. This is particularly important during tripping and setting the well casing. The slower drilling reduces the heating while the greater mud weight increases the down-hole pressure. The zone of decomposition is usually cased with an extra cement collar, the casing being at least 34 cm thick.

Since permafrost provides a low-permeability layer over all the underlying rocks, shallow high pressure water zones may be found beneath it. Water squeezed out as rocks become compacted cannot escape, so unusually high pressures may be encountered immediately below the permafrost zone. The water may have enough pressure to flow out at the surface, in which case it must be allowed to flow until the pressure is reduced or until the water freezes. Relief wells may be necessary and the pressure must be kept low if the well is drilled to greater depths, to avoid casing failures.

SUMP PROBLEMS

Unfortunately, about 20-25% of the wells drilled in northern Canada have experienced problems with sumps (French, 1978, 1980). The first type of problem is underestimating the volume of drilling fluid. An undersized sump may result from misinterpretation of the geology or a poor blast during excavation. Changes in plans for depth of drilling may often be the cause, and at Chads Creek B-64 on Melville Island, three sumps had to be constructed. The second type of problem is due to thawing of ground ice around the sump. At Bent Horn A-02 on Cameron

Island, the drilling rig had to be taken down and evacuated in a hurry during summer and the well was abandoned as the ground thawed and turned into a lake which engulfed the pad. At other times, drilling fluid may leak out of a sump into a stream (e.g. Parsons Lake D-20, Mackenzie Delta), or lead and cause slope failure (e.g., Caribou N-25, Peel Plateau, northern Yukon).

Additional problems with sumps can occur during the restoration process. At Parsons Lake A-44, the freeze-back after infilling created pressures that resulted in a volcano type effect. The result is similar to a pingo. At other times, the fluid drains away after melting the surrounding ground ice, with the result that there is sump subsidence and collapse, e.g. at Beaverhouse Creek N-13. Blasting of the sump may leave too little material for infilling (e.g. Niglintgak M-19 and H-30) or the fluid leaks out prior to freezing (e.g. Parsons Lake D-20). If the ground contains large quantities of massive ice, there may be no overburden available for covering the sump. Sometimes poor choice of site creates problems with the sump, e.g. when the sump is located on the edge of a steep slope which then fails, or in a stream bed which is later eroded by a river. Finally, there may be excessive terrain disturbance (e.g. Bent Horn A-02) due to thermokarst and poor freeze-back. The fluids are often mixed so that they do not freeze at 0°C. Snow must not be used in the cover for it will tend to melt out and cause collapse in summer.

There is now fairly tight regulatory control of drilling on land in North America. The greatest weakness appears to be in deciding what constitutes a toxic fluid. The most common test appears to be to put several fish into the substance and determine the level of toxicity that kills off 50% of the fish in x hours. Here, there has to be a clear distinction between the effects of the concentration of the substances and their chemical composition.

IN SITU GAS HYDRATES

Gas hydrates are inclusion compounds in which small guest molecules of natural gas fit into the structural voids of the ice lattice. Methane, ethane, propane and isobutane can be involved (Davidson, 1973).

Once a possible structure has been found, the additional potential problems of gas hydrates

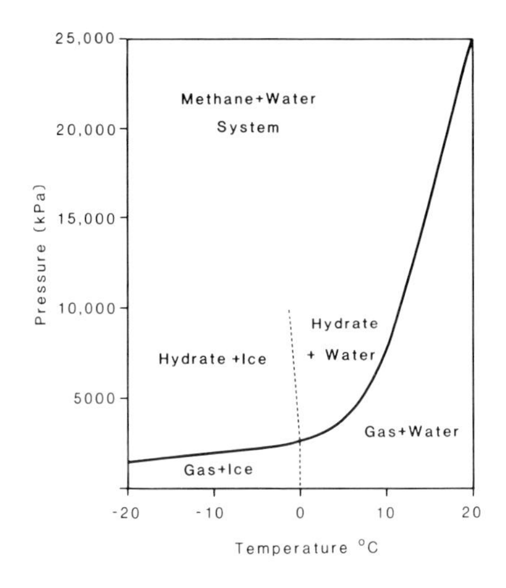

Figure 7.2: Stability of the Methane-Water-Gas Hydrate System (after Davidson et al., 1978, p. 938). Reproduced with the permission of the National Research Council of Canada.

associated with permafrost conditions must be considered. Gas hydrates were first encountered when they caused plugging of pipelines (Hammerschmidt, 1934). The first realisation that they could occur in the ground was by Strizhev in 1946 (Makogan, 1974) and this was proved correct in the Markhinskaya well in Northwest Yakutia in 1963. Signs of possible gas hydrates have since been found in many places at shallower depths in the Arctic (Judge, 1982), as well as in the high pressure - low temperature environments of the ocean depths by the Lamont-Columbia group.

Figure 7.2 shows the pressure-temperature controls on the methane-hydrate system when excess water is present. It is similar in medium sands (Baker, 1974; Stoll, 1974) but hydrates may not occur in clays (Weaver & Stewart, 1982). Similar curves are obtained for the phase systems of other gases. The colder the system, the greater the probability of hydrocarbons being at least partly in the form of gas hydrates in the upper portions of the ground. If decomposition occurs, there is an immediate change in volume as the solid changes partly to gas. Gas-pressure "kicks" are the normal result. If the column of drilling mud is not heavy enough to resist the pressure and/or gas in the mud reduces the weight of the mud, a blowout can occur.

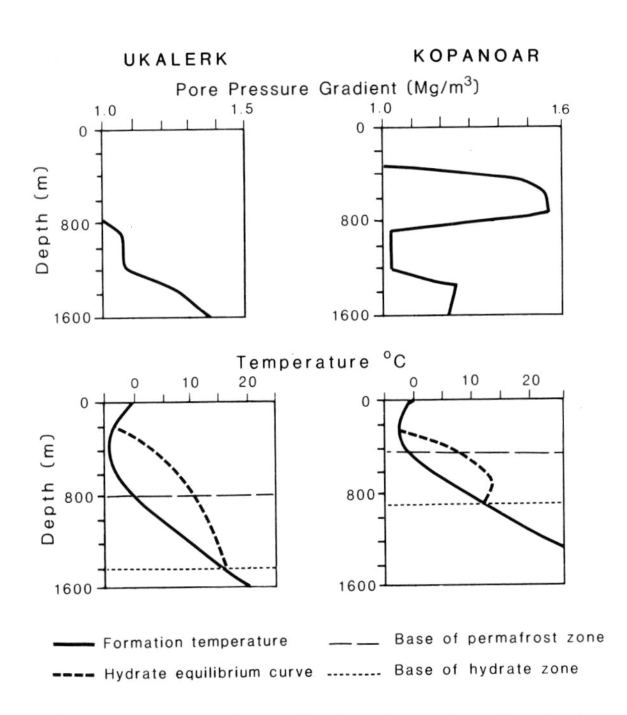

Figure 7.3: Prediction of the *in situ* Hydrate Zone Using Formation Temperatures and Pore Pressure Gradients (after Weaver & Stewart, 1982). Reproduced with the permission of the National Research Council of Canada.

On occasion, the entire drill stem is believed to have been blown out of the ground.

Gas hydrates can be identified during drilling by intense gasification of the drilling mud. This is due to the heat generated by the drilling bit causing thawing of the gas hydrates. This can be overcome by slower penetration rates and by using cold, heavy drilling mud (Goodman & Franklin, 1982). The heavy mud may cause problems but the colder fluid should stop or minimise gasification (Weaver & Stewart, 1982).

Zones of potential gas hydrates can be predicted from geophysical information such as the analysis of seismic data. The best data come from drillstem tests and analysis of wireline and mud-gas logs. Given the composition of the formation gases, the composition of the formation water, the formation pore pressures and formation temperatures, the base of the potential hydrate zone can be predicted quite

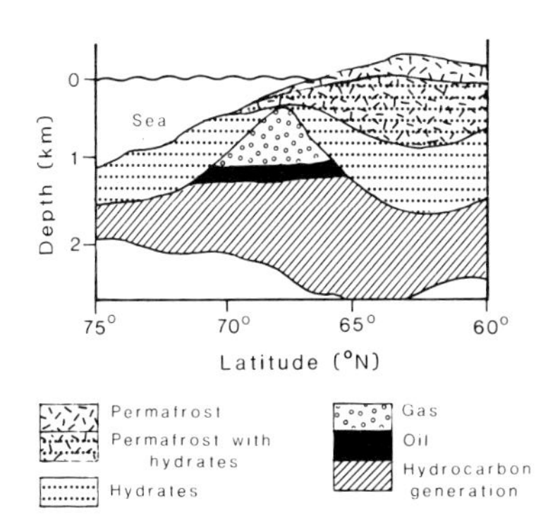

Figure 7.4: Zonation of Permafrost and Hydrates in the Russian Arctic (after Makogan, 1982). Reproduced with the permission of the National Research Council of Canada.

accurately (Fig. 7.3, after Weaver & Stewart, 1982). The composition of the gases determines which phase diagram should be used, while the composition of the formation water determines the conditions of the water-ice phase change.

The base of the hydrate zone roughly parallels the base of the permafrost (Fig. 7.3) but can be appreciably below it. There is a deepening of the base of the hydrate zone across the Arctic in North America. At Prudhoe Bay, hydrates occur in 32 out of 125 wells at a depth between 200 and 290 m (Collett, 1983). In the Beaufort Sea, the base of the zone of *in situ* hydrates varies from 700 m in the west to 1250 m in the east (Weaver & Stewart, 1982) and this depth is probably similar in the Eastern Arctic and Arctic Islands (Judge, 1982). Figure 7.4 shows the probable relationships of the oil, gas, hydrate and permafrost zones in part of Arctic Russia (Makogan, 1982). Gas hydrates do not normally occur above 140 m in depth, nor below about 1860 m (Judge, 1982).

ICE TYPES AT SEA

Before examining the problems of off-shore drilling, it is necessary to explain the nature of the ice cover in the area. The Beaufort Sea may be taken as a convenient and well-documented case. Figure 7.5 shows the main ice types as they are found in winter, while Figure 7.6 shows a cross-section of the ice.

Near shore and adhering to the shore is a region of landfast ice which usually starts to form in

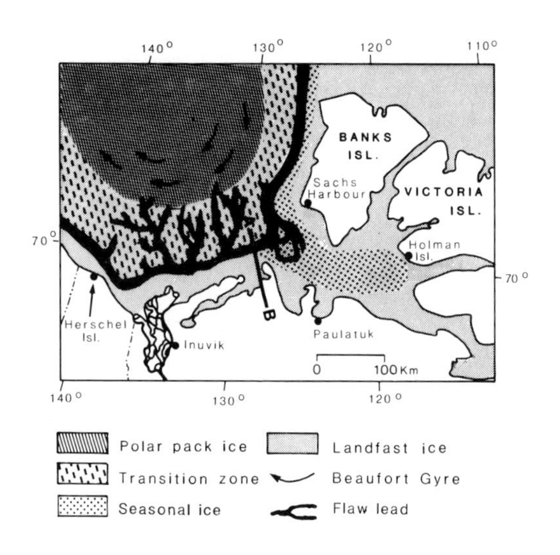

Figure 7.5: Ice Types in the Beaufort Sea Based on Observations During the Period 1973-1980.

October and slowly grows out to 20 m water depth by December. Transition zone ice is mainly seasonal and moves onshore and offshore with the winds, forming heavy ridges. These ridges consist of mounds of broken ice formed by the converging ice sheets along the northern edge of the landfast ice.

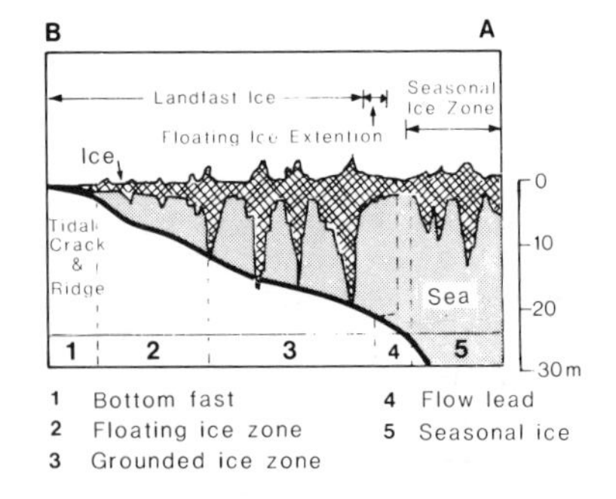

Figure 7.6: Cross-section of Landfast Ice Zone Showing the Grounded Pressure Ridges (after Spedding, 1974; Gladwell, 1976).

Fragments of ice islands or pieces of multiyear ice may become trapped in the landfast ice in the Fall and remain there until the Spring (Cooper, 1975; Ramsier et al., 1975), although this only happens about once in five years (Spedding, 1978). Landfast ice reaches thicknesses of about 2 m by mid-May before melting rapidly in about two months.

Pressure ridges tend to have keels extending below water to about four times the sail height, so that those with sails over 5 m high are generally grounded. These grounded ridges aid in stabilising the landfast ice.

The ridges probably play an important role in absorbing and deflecting the movements of the transition zone and polar pack ice (see Fig. 7.5). Grounded ridges form parallel to the margin of the moving ice and are formed as temporary shear zones.

The shear zone commences at the edge of the fast ice and continues out for 100 km to 200 km. It is very mobile and can move at any time. It is formed predominantly of first year ice, although multi-year ice can be driven into it. The motion of this ice is primarily caused by winds. This can open up or close leads or breaks in the ice up to 10 km wide in 15 hours (Cooper, 1965). The ice remains weak and mobile all winter.

Further offshore is the polar pack zone. This is first year ice that has survived one or more melt seasons, i.e., it is multiyear ice. This ice is generally formed from pressure ridges which, after one or two years, become solid masses of ice and hence a formidable hazard to ice-breaking vessels and fixed structures. The mean multiyear ice thickness is 5-6 m, but many pressure ridges are over 20 m thick. The largest multiyear ridge measured to date was 43 m (R. Pilkington, personal communication, 1985).

Massive "hummock" fields may occur in the pack. They consist of ice floes several km in diameter containing parallel rows of thick multi-year ice ridges. The hummocks form on the west coast of the Arctic Islands, principally off Borden and Prince Patrick Islands. The exact calving rate of hummock-fields is not known.

Occasionally, giant ice islands are found within the polar pack ice. These ice islands may be up to 250 square km in area and 20-50 m thick and originate as blocks of shelf-ice or floating glaciers on the northwest coast of Ellesmere Island (Hattersley-Smith, 1963; Holdsworth, 1971; Jeffries, 1983, 1984).

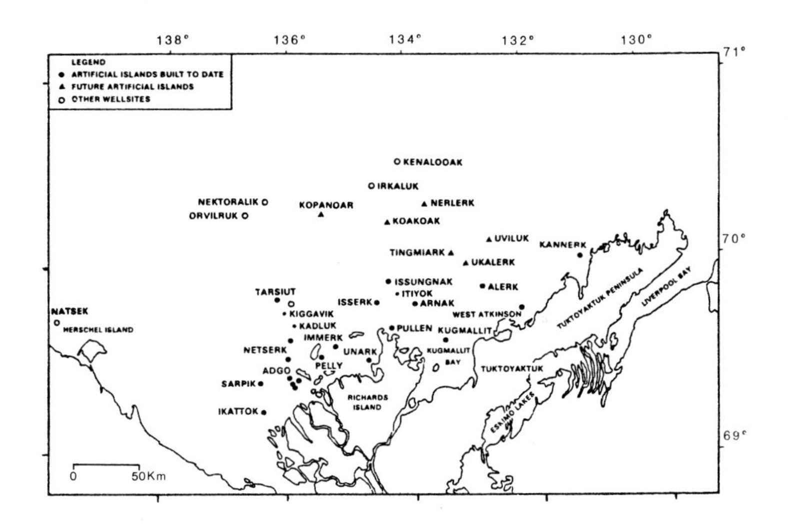

Figure 7.7: Wellsites in the Offshore Beaufort Sea in 1982 (after Anon., 1982).

Frequency of calving of these ice islands is not well known but it appears that the ice on Ellesmere Island regenerates quickly (Jeffries, 1984).

Great grooves or scours are abundant on many parts of the floor of the Beaufort Sea. Some appear quite fresh and others are quite old, but scours are difficult to date. They also occur in the zone of landfast ice where they are believed to be the result of grounding of pressure ridges.

The position of the boundaries of the ice types varies from year to year (Sobzak, 1977; Spedding, 1974, 1975, 1979). Similarly, studies of the movements of the ice features in the Beaufort gyre show large variations. The ocean currents also show considerable variability (Bernstein, 1972). Hence, it is extremely difficult to determine the probability of occurrence of extreme conditions at an offshore structure (Spedding, 1978).

From this discussion, it will be seen that the zone of landfast ice is the safest for offshore structures, while the risk of damage by ice forces increases progressively offshore. In the zone of the pack and polar pack ice, the design of the structure must allow for almost continuous movement and the possible impact by a thick, large multiyear feature.

OFFSHORE DRILLING

Offshore drilling in the Beaufort Sea (Fig. 7.7) has been affected by history and economics. Esso Resources Canada Ltd. purchased the exploration rights to most of the sea floor out to 20 m water depth, Gulf Canada Ltd. the sea floor from 20-30 m depth and Dome Petroleum Ltd. the land beyond. To date, only exploratory and step-out wells have been drilled offshore in the Beaufort Sea. The nature and amount of eventual production will depend on the results of the present drilling programme and economics.

Since 1972, Esso has built 22 sand-filled artificial islands to drill on their lease, since this is the most economical and safe method of drilling in shallow water (down to 19 m). Recently, Esso has been using a caisson-retained island (CRI) system for the deeper water.

In 1976, Dome started drilling in the ice-infested deeper waters during the summer months, using four reinforced vessels. The first caisson-retained island using the shear core principle was constructed in 1981 in 23 m of water at Tarsiut. Subsequently in 1982, they built a second water-filled caisson system out of an old supertanker (the "SSDC").

In 1983, Gulf started drilling in the area with a conical-shaped moored drillship and, in 1984, with their Mobile Arctic Caisson which sits on the sea bed in 19 m of water.

In isolated bays in the Sverdrup Basin where the ice is stationary, Panarctic Oils Ltd. have drilled off thickened ice islands in very shallow water (about 2 m deep).

Artificial Islands

These are discussed in the Environmental Impact Statement for the Beaufort Sea (Anon., 1982, vol. 2), while Maxwell et al. (1983) compare them to concrete and steel caissons. The first island was Imerk B-48, constructed by Imperial Oil in 1972-73 in the Beaufort Sea. Figure 7.7 shows the location of the 22 artificial islands constructed during the next ten years. They are also used as bases for concrete and steel caissons. All but one have been expendable and are allowed to be eroded by wave and ice action.

The first islands were small and built in summer in water shallower than 3.5 m to keep the

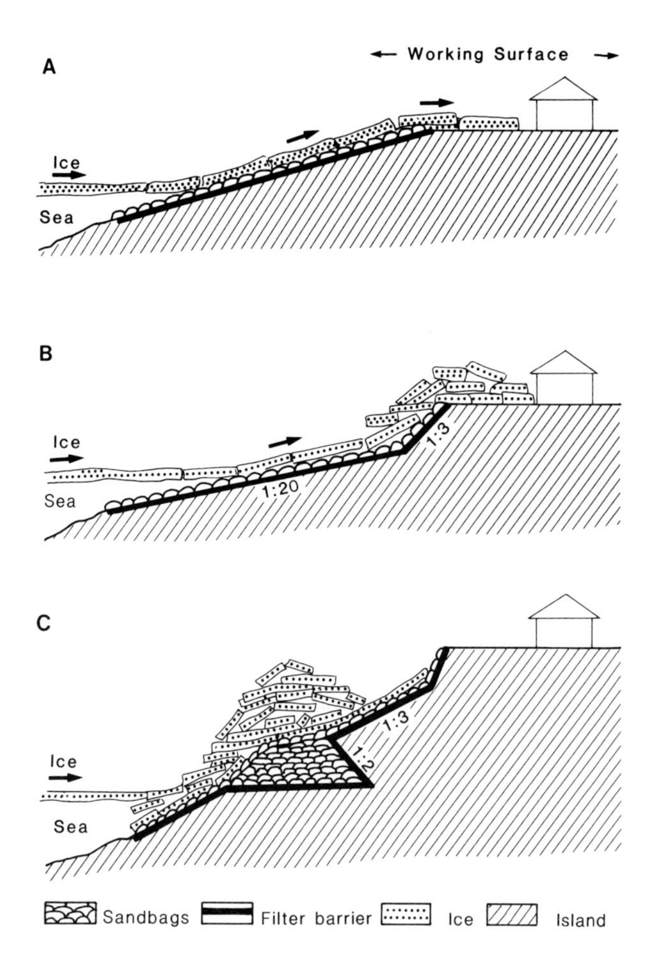

Figure 7.8: Effects of Ice Movement on Shorelines of Artificial Islands of Various Designs (partly after Croasdale et al., 1978). A, normal sacrificial beach; B, modified sacrificial beach; C, maximum beach protection for deeper water.

quantity of material needed to reasonable amounts. The sand and silt were either dredged from the sea floor by specially designed dredges (12 X 10^6 m^3 annual dredging capacity) or transported by hopper barges. The good quality sand material required for the island had to come from relatively far away. Such material must be used to avoid slumping of the island. Speed of construction decreases rapidly with increasing distance, e.g., a given fleet of dredges could only supply half the material from 30 km that they could from 10 km from the island; hence, quantities of material were minimised. The exploratory islands are designed to withstand estimated 20-year ice events and 20-year extreme wave heights. Production islands could be built much larger to resist all ice pressures and a 1000-year extreme ice event but it would be very expensive. These exploratory islands had an outer subaqueous slope of about 1:15 and were liable to potential slope failure due to rotational slippage (Maxwell et al., 1983). While they withstood landfast seasonal ice satisfactorily, they may not withstand the impact of thicker multiyear ice or ice islands, nor may they be stable in moving first year ice up to 2 m thick (Dingle, 1981; Pilkington et al., 1983).

Since the cost of these islands is proportional to the cube of the water depth, the Issungnak islands in 19 m of water requiring 4.1 X 10^6 m^3 fill are at the economic limit for exploratory islands. Permanent production islands in this depth of water would require seven times as much fill, as well as extra defences against waves and ice.

Winter construction was practised for seven out of ten artificial islands constructed by 1983 in shallow water near shore, due to environmental problems during the whaling season in summer. All the material is transported by ice road from land and dumped through a hole in the ice (Tart, 1983). Gravel is normally used and tends to form steep slopes up to 1:0.5. Additional material must be placed around the island as necessary so as to allow for the 1:3 final slope and an adequately-sized working platform. This method originated in Canada in 1973 but is no longer used there.

In early winter, movement of thin landfast ice up to 60 cm thick causes riding up on the shores of the island (Fig. 7.8A). The first designs involved the use of wide "sacrificial beaches" of sandbags on a 1:12 slope laid over a filter barrier. These work well inshore, dissipating more energy on the wide, shallow slope, but with a simple slope, slabs of broken ice can be pushed onto and across the working surface (Croasdale et al., 1978; Croasdale & Marcellus, 1978; Frederking & Wright, 1980; Kovacs & Sodhi, 1979; Kry, 1977, 1980; Strilchuk, 1977). By putting a sharp break in slope at the top of the beach (Fig. 7.8B), better protection could be achieved, although the sandbags may have to be replaced regularly. This design copes adequately with wave action. Even then the ice may jam at the break in slope and be overridden to form a pile spilling over onto the working platform if there are large amounts of ice movement. Thus, these sacrificial beaches can only be used close to shore in very shallow water. In deeper water, greater freeboard and a wider beach with a more complex slope (Fig. 7.8C) ensure that the ice piles up on the shore.

The piling up of the ice has two effects. Firstly, it acts as additional protection for the island against overriding by later stronger ice sheets. Secondly, it enlarges the effective size of the island so that it can survive rather greater forces.

Artificially Thickened Ice Platforms

In the high Arctic, isolated, deep-water bays occur where there is minimal ice movement. The ice is artificially thickened to 6.3 m in a zone 200 m in diameter, so that a drilling rig can be erected on the ice platform (Kivisild & Iyer, 1978). This has been successfully used south of Ellef Rignes Island. Drilling must be completed and the rig removed between November and July. The well head is set on the sea floor.

Drillships

In 1976, three reinforced drillships commenced work in the Beaufort Sea, with a fourth being added one year later. These vessels have successfully drilled in 23-68 m of water from late break-up to early freeze-up. Given light ice conditions, the season can be extended into November using specially built ice-breaking support ships. The Tarsiut discovery well was drilled in this way. Drillships have the potential to work in water depths down to 200 m (Maxwell et al., 1983).

Concrete and Steel Caissons and Structures

Hnatiuk and Wright (1984) have discussed the five main types of concrete and steel caissons and drilling units and problems encountered with them. The five main types are: shallow caisson islands, single steel drilling caissons, steel caisson-retained islands, conical drilling units and mobile Arctic caissons. These systems are designed to operate in 15-40 m of water, either sitting on the sea floor or placed on berms of various heights,

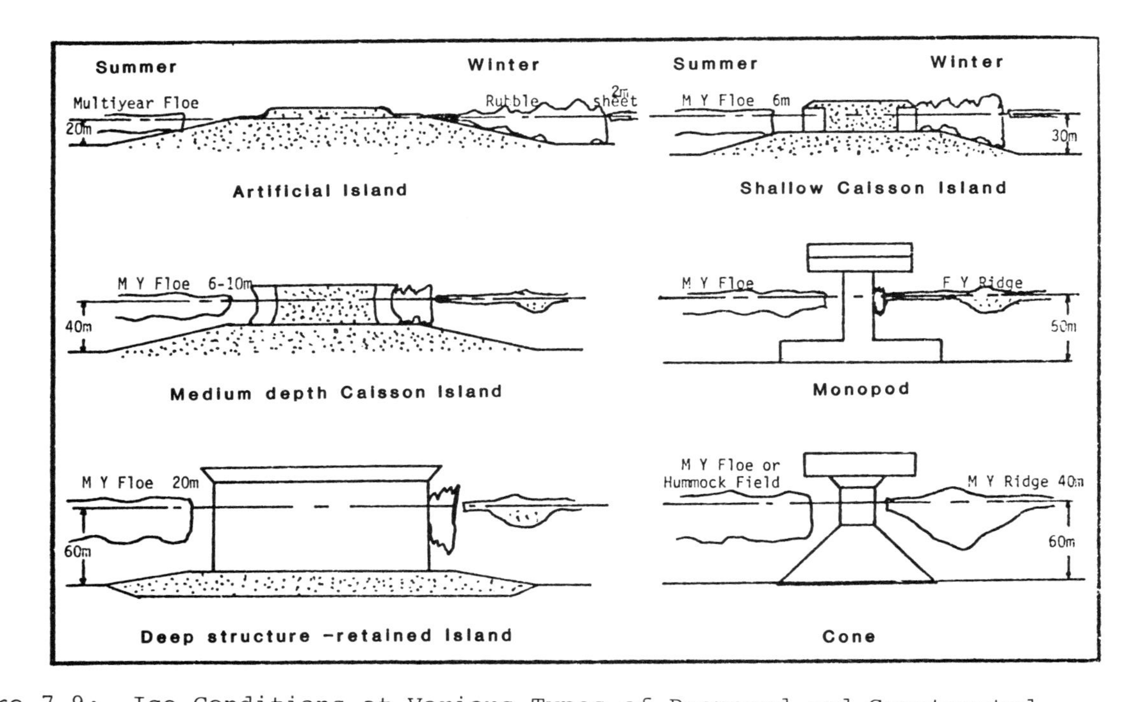

Figure 7.9: Ice Conditions at Various Types of Proposed and Constructed Structures in Different Water Depths (after Maxwell et al., 1983). The deep structure-retained island is a steel or concrete caisson island.

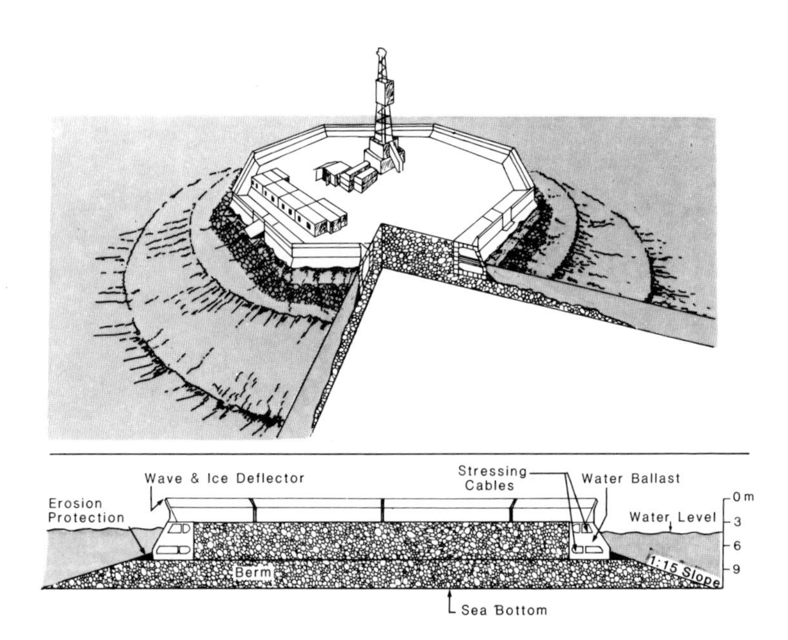

Figure 7.10: The Caisson-retained Island Used in the Beaufort Sea by Esso Resources Ltd. (Dingle, 1981).

thus reducing the amount of fill needed. Figure 7.9 shows how the various types of ice interact with the different structures (after Maxwell et al., 1983, Fig. 7

Tarsiut was the first caisson-retained island and was constructed in 1981 in 22 m of water. The berm was built to form a suitable platform within 6 m of the water surface. Four concrete caissons, 11 m high, 80 m long and 15 m wide were placed on the platform, steel doors closed the gaps and then the inside was filled with sand and gravel prior to freeze-up to produce a shallow caisson island (Fig. 7.10).

The single steel drilling caisson consisted of a segment of an old oil tanker. It was reinforced with a double hull with concrete between the shells to withstand ice pressures and drilling equipment was installed on its decks. It was then ballasted on to a subsea berm built to within 9 m of the water surface to provide a medium depth caisson island (Fig. 7.9). The Canmar SSDC successfully permitted drilling of wells at two locations in moving pack ice in 1982 and 1983.

The Esso Caisson-Retained Island (CRI)(Fig. 7.10) can be used in up to 20 m of water. It consists of floating caisson segments held together with heavy steel cables ballasted onto a suitable berm (Dingle, 1981). The interior is then filled with dredged sand to provide resistance to sliding caused by ice pressures, to form a shallow to medium depth structure (Fig. 7.9). After sand filling, the drill rig and accommodation must be installed. The Gulf Molikpaq (operated by Beaudril Ltd.) is a single mobile Arctic caisson which can drill in 21 m of water and can be moved to a new location each summer (Hnatiuk & Felzien, 1985). Like the Esso CRI, it consists of an outer steel caisson filled with sand but the drill rig is permanently installed on the deck. The advantage of these designs is that they offer the least resistance to ice and the cone breaks up the ice mass pressing against the vessel.

The Gulf Canada Ltd. drill vessel Kulluk is a free-floating vessel, anchored in water with a 12 point mooring system (Hnatiuk & Wright, 1984). Its major advantage is that it needs no sand fill nor a berm to sit on and is, therefore, far cheaper.

PRODUCTION WELLS

Oil Wells

In the Prudhoe Bay field of Alaska, oil reaches the ground surface at a temperature of about 80°C. Whether it occurs on land or in the sea, this causes a considerable thermal disturbance that must be allowed for around the well casing.

The axial stresses applied to the well casing due to thawing are shown in Figure 7.11 (after Goodman et al., 1982). These forces are greatest at the base of the permafrost and put the casing under tremendous strain. Thaw discontinuities and lithological changes cause fluctuating pressures on the casings within the permafrost zone. Special casings and materials may be necessary (Goodman, 1978a) and surface-controlled subsurface safety valves are placed below the permafrost base (Goodman, 1978b). Tubular expansion joints which can accommodate tubular length changes of up to 6 m have been developed for the casing above the producing zone. Changes of 5 m are commonplace during injection of cold treating fluid and movement is essential if the casing is to survive. By using double casings, insulating packer fluid and circulating collars, it is possible to minimize the heating effects (Goodman, 1978c; Goodman & Mitchell, 1978). Insulated tubing

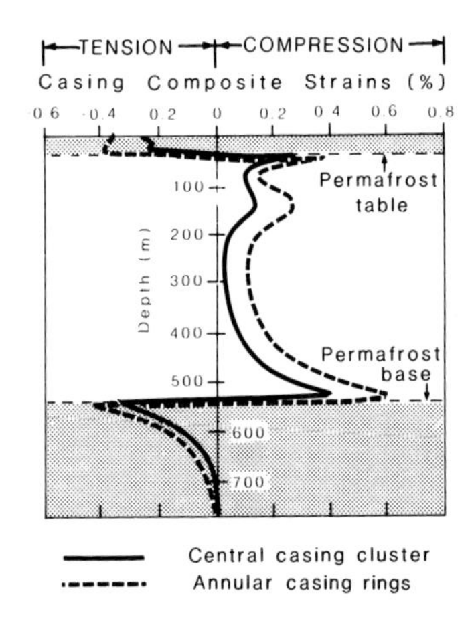

Figure 7.11: Theoretical Maximum Magnitudes for Casing Composite Strains for Multiple Wells Either Clustered or Placed in Annular Rings (after Goodman et al., 1982). In layered materials, these strains may increase by a factor of four. Reproduced with the permission of the National Research Council of Canada.

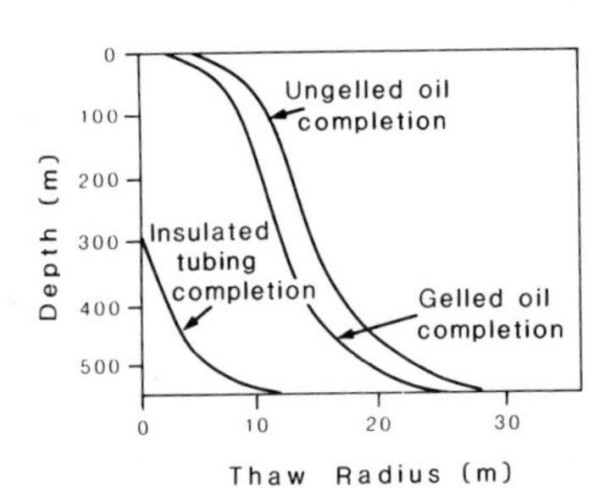

Figure 7.12: Comparison of Calculated Thaw Radii in the Permafrost Zone for 20 Years of Production of Oil at 80°C at Prudhoe Bay, Using Different Thermal Protection Methods (after Goodman, 1978c). Note the increase in thaw radius with depth where the oil is warmer.

will eliminate the thaw in the upper half of the permafrost (Fig. 7.12), while gelled oil packer fluid noticeably decreases the thaw radius. The increase in width of the thaw zone with depth is due to the higher oil temperatures at greater depth. In theory, refrigerated conductor casing and heat pipes could be used to provide greater thermal stability but they have yet to be used in production wells, probably due to their cost.

If the temperatures around the casing are not prevented from rising to the temperature of the oil or gas being produced (80°C at Prudhoe Bay), the pressure of gas produced by melting gas hydrates will become enormous and can collapse the casing (Fig. 7.2).

When production ceases temporarily or permanently, freeze-back may cause extra pressures to be applied to the casings (Goodman, 1977). The annular ring of thawed material requires more than its original volume when it refreezes. The resultant pressures are transmitted towards the borehole. Given an initial thaw radius of 0.75 m at Prudhoe Bay, theory suggests that pressures may reach 1.6 times the overburden pressure at any given depth. It must also be realized that when multiple wells are drilled from an artificial island, the cumulative effects may be much higher.

PRODUCTION OF GAS FROM GAS HYDRATES IN WELLS

It has been estimated that 10^{13} m^3 of natural gas is present in hydrate form in the USSR (Cherskii & Makogan, 1970; Trofimuk et al., 1980). Davidson et al. (1978) have suggested that 10^{11} m^3 of gas as gas hydrates may be present in the Mackenzie Delta. The total proven reserves in this form may increase as exploration proceeds. This represents a tremendous potential source of gas which will eventually be needed by the main population centres and by industry.

Safe and inexpensive production from gas hydrates is at present a problem. Controlled, slow decomposition of the gas hydrates would be required (see Fig. 7.2) and it will be necessary to turn the flow on and off without creating dangerous pressures.

Supplying heat to the ground will cause decomposition of gas hydrates but the heat required will be 20% greater than for melting the same volume of pure ice. If the reservoir pressure is decreased (as is usually the case in a producing well), melting will also occur. Brine injection is reasonably cheap and will cause production by melting the hydrate at temperatures between -2.5°C and 0°C (Kobayashi et al., 1951).

Methanol has been used successfully for avoiding or clearing blockages of gas pipelines by gas hydrates (Hammerschmidt, 1934), although drying the gas prior to transport or using higher pressures in the pipeline will avoid the problem. However, in production zones in wells, there has been mixed success in using ethanol and methanol. In the USSR, Cherskii and Bondarev (1972), Makogan (1974), and Kolodeznyi and Arshinov (1970) all report success in stimulating production, whereas Bily and Dick (1974), working in the Mackenzie Delta, were apparently unsuccessful in achieving commercially acceptable production. This is probably because the type 1 hydrate contains a 6:1 ratio of water to methane molecules; this means that large quantities of alcohol are needed to free a relatively small quantity of methane.

Finally, steam and hot-water injection have been suggested as other possible methods (McGuire, 1982) but these have yet to be established as effective and safe methods of production.

OIL AND GAS COLLECTION AND MOVEMENT

These raise a number of questions. Are the wells onshore or offshore? Do they produce large quantities of oil or gas? Gas may be liquified and may need a separate terminal. If the production is offshore, will it be collected there at a central point or will it be piped to the shore? Will the product need processing in the field? Will it be shipped by pipeline or tanker?

The two main methods currently in use for the transportation of oil and gas are tankers and pipelines and both are proposed for use in North America (Fig. 7.13). In Siberia, the main method of transport is by pipeline (Fig. 7.14) and many of the pipelines outside the permafrost zone consist of multiple large diameter (greater than 1 m) pipes.

TANKER TRANSPORT IN THE ARCTIC

One of the competing methods for exporting oil or liquefied gas from the Beaufort Sea is by tanker. The proposed route to the east is through the Prince of Wales Strait (Fig. 7.13), Viscount Melville Sound,

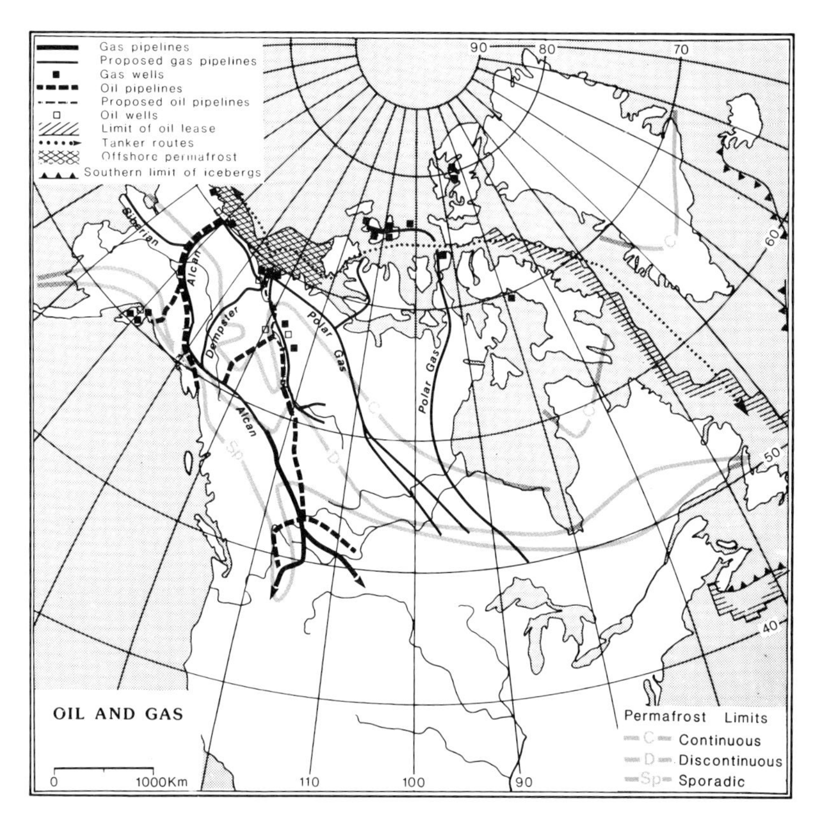

Figure 7.13: Proposed Tanker Routes, Limits of Oil Leases, and Proposed and Completed Pipelines in Northern North America.

Barrow Strait, Lancaster Sound and the Davis Strait. Tankers could also go via the Bering Strait to Japan or the west coast of North America.

This method would require offshore terminals connected to wells via pipelines. Onshore wells could also be connected to the system. The bottom of the Beaufort Sea is a submerged valley and plain area extending about 150 km to the north of Tuktoyaktuk. The sea floor is dotted with submarine mounds, some reaching within 18 m of the water surface. As such, they pose a navigation hazard to large tankers.

Tankers would have to pass through very heavy ice while travelling between the Arctic Islands and it has been found that they can do this

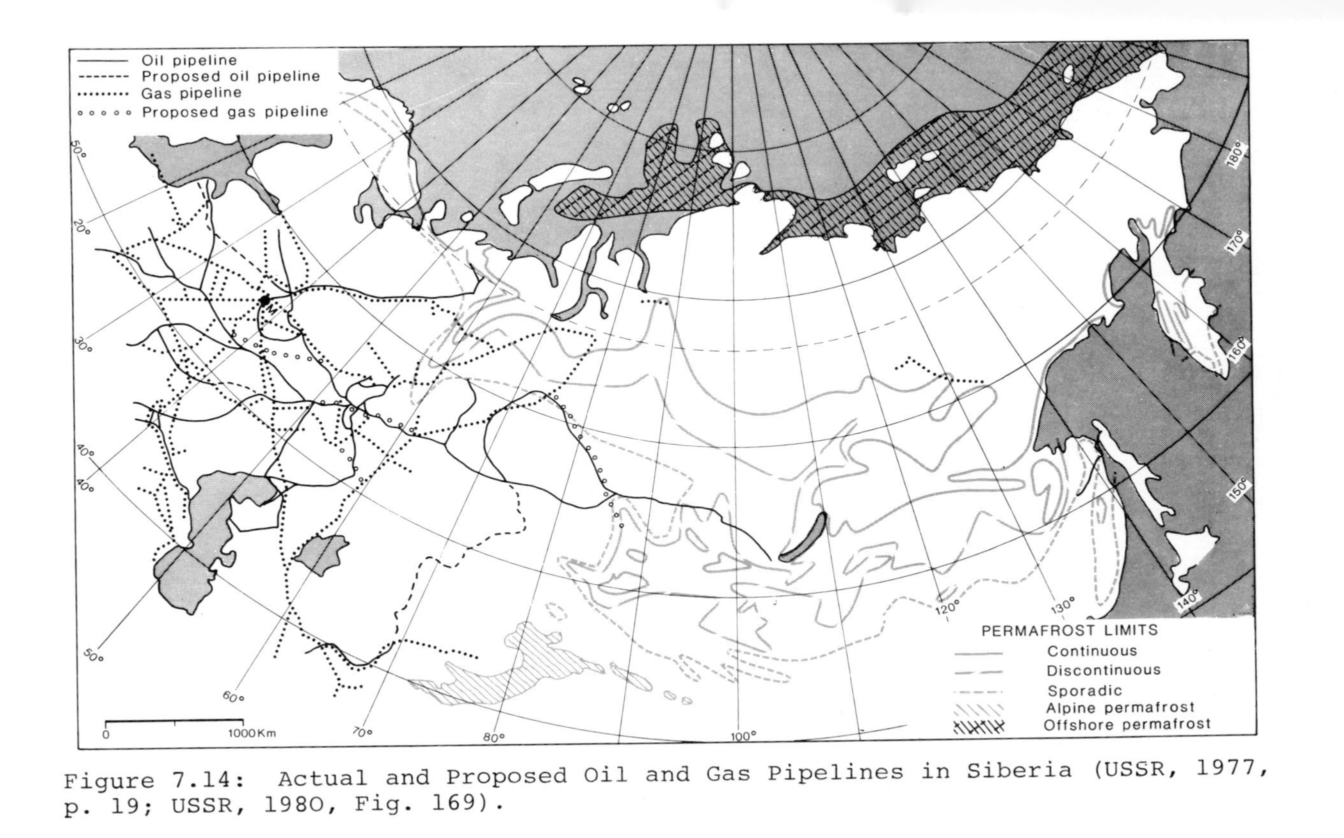

Figure 7.14: Actual and Proposed Oil and Gas Pipelines in Siberia (USSR, 1977, p. 19; USSR, 1980, Fig. 169).

satisfactorily if suitably constructed and powered. They need 20 m draft and also good maps and information about ice conditions. The SS Manhattan was the first tanker to try the route in 1970, escorted by an icebreaker. These vessels succeeded in passing through the area in summer. There is always the risk of getting frozen in or damaged or of running aground, but this is less likely with the class of vessels proposed for commercial use in these areas.

There is considerable debate over the use of tankers in the Arctic. The ecosystem is very fragile, with the most productive parts of the ecosystem being located at points where the sea does not freeze in winter (e.g. Baffin Bay, shear zones, and Durdos polynia). The tracks of the ships will make the ice unsafe for one or two hours after its passage, which may cause difficulties for animals (e.g. Peary caribou and muskox) and for hunters. Dome Petroleum claims and has demonstrated that heavy sleds can pass over the refrozen tracks after two hours, but this undoubtedly will vary from place to place and with the season of the year. The ships would increase the number of breaks or natural leads in the ice caused by winds, although the risks would be similar to those existing now.

The main reason that tanker transport is seriously considered is that as soon as terminals and a tanker are built, they can go into service, thus providing an immediate return on the investment. Pipelines can only go into use when the entire system is ready for operation. Hence, more capital is tied up far longer without producing any return on investment.

The risk of an oil spill is real, and with larger tankers, the problem may become worse. Hence, there is considerable debate over the potential use of tankers. However, tanker designs indicate that the vessels will be stronger than any constructed to date, just to withstand the ice conditions. The vessels will be adequately powered to move continuously through the ice, otherwise the operation would not be economical. Also, regulations will have to call for built-in redundancy in power systems, steering, radar, depth sounders and navigation systems. Numerous safety systems are being considered which reduce the chance of an oil spill in the event that the vessel's outer shell is ruptured.

PIPELINES

Oil leaves the ground at 80°C at Prudhoe Bay; it must be pumped at 60-65°C because it is very viscous at 0°C and will not flow through the pipe. However, some other crude oils, e.g., at Norman Wells, are as mobile as diesel oil.

Gas can be moved at any convenient temperature if the pressure gradient is high enough. Thus, the suggested maximum operating pressure for the proposed Arctic Gas pipeline was 11.58 Mpa. This is equivalent to the pressure at about 1100 m under the sea. The gas itself is potentially highly explosive, as was demonstrated by a break in a lower pressure pipeline at Brooks, Alberta. If there is a leak, and there are sparks, the gas explodes and burns until all the gas in the pipe is exhausted.

At whatever temperature the fluid is moved, there is a potential for trouble somewhere along the route from the permafrost area to the market. A warm pipe will tend to cause thawing and settlement of the soil over permafrost, whereas a cold pipe tends to cause freezing of the ground with attendant heaving in frost-susceptible materials (Jahns, 1984). Even cooling of frozen ground may cause additional build-up of ice around the pipeline and risk of vertical displacement. Special designs are needed on the seafloor (Jahns, 1984a).

For all pipelines it must be remembered that energy is required to move fluids uphill. The shortest route is not necessarily the most efficient. In North America it is usual to run a pipeline around an upstanding topographic feature. Again, in general, fewer problems are encountered in permafrost by building pipelines on the wet lowland than on better-drained terraces. Finally, it should be noted that the period of use of the facility is a fundamental part of the design. Even if the design is a good one and the facility well-constructed, the structure will need to be largely replaced at the end of the design life, so it is essential that all the fluids to be produced from the area are moved by then.

OIL PIPELINES

The first oil pipeline built on permafrost was a small diameter pipeline built to bypass the St. Charles rapids on the Great Bear river in the 1930 s. It supplied fuel from Norman Wells to the uranium mine on the shores of Great Bear lake. The next appears to have been the Canol pipeline, built from the Norman Wells oilfield to Whitehorse in 1943. It filled a strategic need to provide oil for a small refinery along the Alaska Highway, but it was abandoned a year later after a burst following thaw subsidence at a time when the threat of a Japanese invasion of North America abated. It was a small diameter pipe (15 cm and 10 cm) laid directly on the ground alongside the service road. A second similar (20 cm) pipeline was laid in the same manner from the north end of the inland passage at Haines, through Haines Junction to Fairbanks. Locally the pipeline was buried with as much as 50 cm of cover. This mode of construction caused relatively few problems apart from the buoyancy of the pipe in the Slims River delta at Kluane Lake. Subsequently, tanker trucks took over the supply of fuel oil and gasoline and the Canol pipeline was abandoned.

The largest oil pipeline project in North America is the Alyeska Pipeline from Prudhoe Bay on the Arctic coast to Valdez on the Gulf of Alaska. From there oil is shipped by tanker to the west coast of the contiguous United States and to Japan. Had tankers been used all the way, it is probable that they could only have moved oil for a few months each year with the technology available at that time. Since a substantial oil field had been discovered by 1968, it was decided a pipeline should be constructed. The larger the oilfield, the more economical it is to use a pipeline. The method of construction and maintenance of the pipeline will be discussed below.

The use of pipelines cannot be totally avoided, even in North America. They must be used to collect the oil from the producing wells and take it to a central shipping terminal, even when using tankers. The basic rules learnt in constructing the Alyeska pipeline must be applied to all oil pipelines constructed on permafrost terrain because a ruptured pipe can create a great amount of environmental damage in a short time. Panarctic Oils have completed a pipeline along the sea floor from an offshore well near Drake Point.

Alyeska Pipeline

This was built between 1969 and 1977 and extends for 1,300 km (Fig. 7.12). There was an eightfold increase in costs between the start of construction and completion, the final cost being over $7 billion.

The pipe itself has a diameter of 1.22 m with a wall thickness of 1.3 cm. Four thousand drillholes

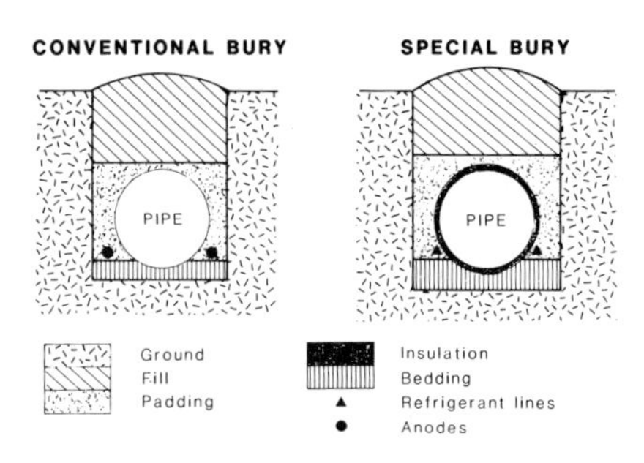

Figure 7.14: Types of Burial Used for the Alyeska Pipeline (after Alyeska Pipeline Service Company, 1977).

were made in 1968 in order to delineate the areas of icy permafrost and a further 2,000 were made in critical areas during construction. The actual trench in which the pipe was laid was examined during construction to identify the nature of the material along the right-of-way, so that the design could be modified as necessary. As a result, only three major mistakes were made in the buried section which resulted in spills. A monitoring programme was established to avoid further problems. Descriptions of the project can be found in Allan (1977), Alyeska Pipeline Service Company (1977), Roscow (1977), P. J. Williams (1979) and Metz et al. (1982).

Design Parameters

The pipeline was constructed over permafrost for 70% of its route and was designed to be elevated above ground on vertical support members (VSM's) wherever permafrost was present in the ground and the soils showed thaw-sensitive properties. Where the soils were thaw-stable or where permafrost was absent, the pipeline was buried (Plate XIII; Fig. 7.14).

Typical thaw-stable materials were dense, frozen, clean sand and gravels and frozen, competent rocks. However, often the surficial materials were ice-rich silts and clays with up to 70% ice. The thaw-stable sands and gravels were defined as having less than 6% water by weight, passing the ASTM number 200 sieve (0.074 mm diameter) and having a dry density greater than 1.92 g/cc.

Plate XIII: Thawed Snow Above Buried Section of the Alyeska Pipeline, 70 km North of Valdez, Alaska.

Plate XIV: Zig-zag Arrangement of the Alyeska Pipeline to Allow for Thermal Expansion and Contraction.

ANCHORED SUPPORT

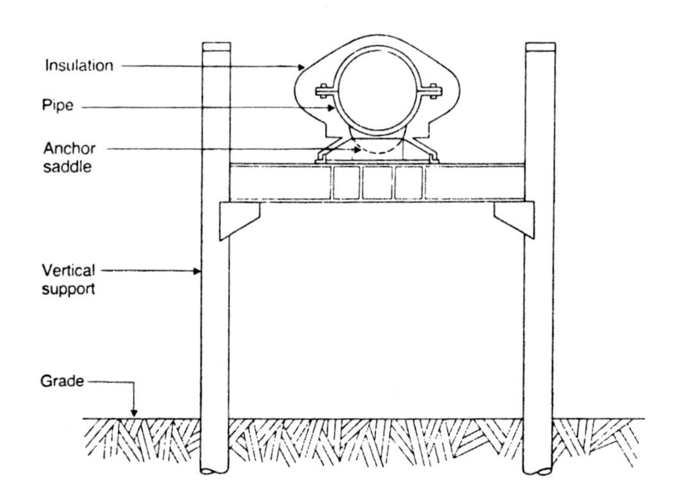

NORMAL SUPPORT

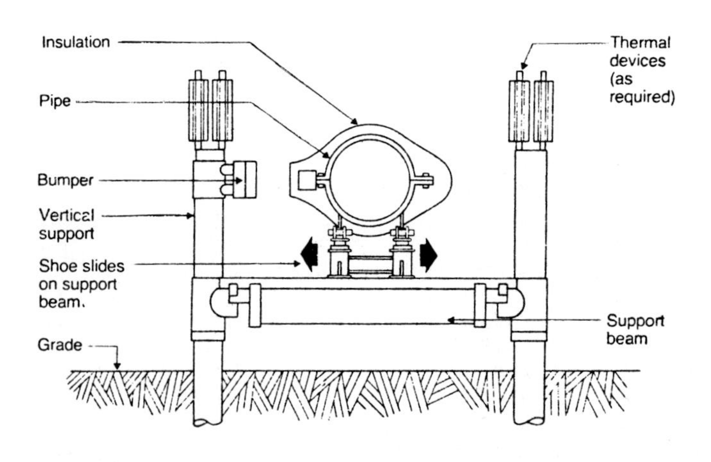

Figure 7.15: Types of Vertical Support Members Used for the Elevated Sections of the Alyeska Pipeline (after Alyeska Pipeline Service Company, 1977).

Above ground, the pipe was covered with 10 cm of resin-impregnated fibreglass insulation inside a galvanized steel jacket. This keeps the oil warm during unexpected shut-downs and helps to avoid problems.

The pipeline is supported on VSM platforms 22 and 25 m apart (Plate XIV). On the normal VSM, the pipe is clamped to the cross beam, which is usually supported on four piles (Fig. 7.15). Thermal piles (Fig. 4.14) in pairs are used wherever there is a risk of thawing the underlying permafrost (Plate XV). The pipes were placed in the drilled holes, which were then backfilled with a sand-cement slurry. Mechanical refrigeration was used to refreeze the thawing ground and the slurry and then heat pipes were installed to replace the mechanical refrigeration system. There were very few problems, with only 0.6% of the 122,000 heat pipes settling more than 9 cm. Some problems were encountered with blocking of thermal piles, which then had to be replaced. Corrosion sometimes produced hydrogen inside the thermal piles, which stopped heat radiation into the atmosphere in winter. By using two thermal piles, if one failed, the other would perform the cooling alone until the defective pile was replaced. Infrared scanning of the thermal piles in winter from a helicopter shows the defective piles as being dark. The piles are designed to have a 30-year lifetime, whereas the pipeline is designed to perform within acceptable settlement limits for 25 years. After that, the pipeline should be rebuilt, if it is still needed.

To compensate for expansion or contraction of the pipe and to allow for earthquakes, the line was built in a flexible zig-zag configuration, which converts the change in length into a sideways motion (Plate XV). On the supports near the bends, the pipe is mounted in a sliding shoe which moves as necessary (Fig. 7.15). The anchored supports were placed at every 245-550 m. At the Denali Fault, special long-width support beams were installed to permit up to 6 m of lateral motion plus up to 1.6 m of vertical displacement.

A total of 142 valves were installed to limit the amount of oil spilled in any accident to 50,000 barrels, but in most sections of the pipeline, the maximum would be 13,000 barrels. Around the Denali Fault, the valves were placed so as to limit the spills to 5,000 barrels of oil. A cleaning device can be slid along the pipe to dislodge any sludge.

Plate XV: Details of Dual Thermal Piles Used as Supports for the Alyeska Pipeline in Unstable Permafrost near Fairbanks, Alaska.

It is moved by the flow of oil, i.e., at between 6.4 and 11.3 km/h. The pipeline delivers almost 200 million litres of oil per day to Valdez.

For limited sections, it was necessary to bury the pipe in thaw-sensitive soils, e.g. at river crossings. In these sections, the pipe was buried in an insulated box (Fig. 7.14) with mechanical refrigeration lines. Some problems have been encountered with failure of the waterproof barrier, which allows water to enter the insulation. The latter then becomes ineffective and the pipes start to corrode due to failure of the coating. These sections will

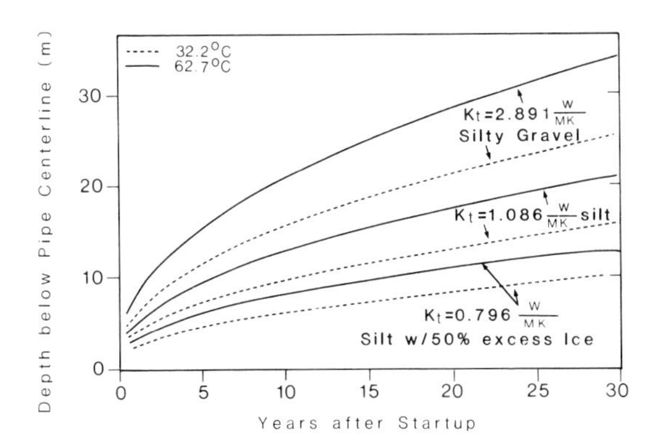

Figure 7.16: Probable Depth of Thaw Below the Alyeska Pipeline as a Function of Time, Temperature of the Oil and Thermal Conductivity of the Ground (Kt) (after E. Johnson, 1981).

have to be re-insulated, which is very expensive, especially in wet soils and under buildings. The pump stations are being tackled first to avoid failure of the structures due to thawing of the underlying ground. Polystyrene is being used instead of the original polyurethane, together with new coating methods. The polyurethane lasted less than three years as an effective insulation.

Figure 7.16 shows the probable depth of thaw below the pipeline as a function of time that was encountered in the design. The progressive thawing is readily apparent and is most marked in coarse-textured sediments. Failure of the insulation in the box means that considerably increased thawing will take place even if the amount of mechanical refrigeration is increased.

Construction Methods

The construction was carried out in winter to minimize the environmental damage and to provide maximum strength as soon as possible after construction. In the most northerly 200 km, a 25 cm natural gas line was added during the winters of 1975-1976 and 1976-1977 to provide fuel for pump stations #2, 3, and 4.

Since there was no access along the part of the route north of Goldstream, a haulroad had to be built to carry the pipe and equipment (Metz, 1984). This was only open to construction traffic, which made construction easier. It was built of 1.5 m of

Plate XVI: Cooling Vanes (Radiator) of a Passive, Two-phase Thermal Pile, Alaska.

thaw-stable gravel laid on the ground in winter using 5-10 cm board insulation where necessary to reduce the thickness of fill required to avoid thermal problems. Revegetation of the disturbed area by re-seeding was commenced immediately (Johnson, 1981). Two roads were constructed, one following the general route for hauling freight and one following alongside the pipeline itself. The latter was restricted to service vehicles to reduce the risks of collision between vehicles and VSMs.

Construction of the pipeline was carried out on an insulated work pad in areas with permafrost. Where available, snow was used as part of the work pad. The use of convection-cooled thermal VSMs (Plates

XV and XVI) meant that the buried portion of the piles for the above-ground section of pipeline could be reduced to about 8 m length. In extreme cases, they still had to extend to 20 m depth to prevent thawing of the ground. Where snow was not used as a work pad, the ground had to be revegetated as quickly as possible.

The gas pipeline was added afterwards by building a new work pad and then trenching through it. In places, snow had to be manufactured to augment the local precipitation, so as to provide an adequate, temporary workpad.

Initially, the pipeline was planned to pass under rivers. In practice, this can create considerable difficulties because it must be placed at least 5 m below the bed of the river to avoid the effect of scour during times of flood. The Tarina crossing was constructed in this way using a 22 cm coating of concrete to protect the pipe in case it became exposed and to give it negative buoyancy.

The beds of the rivers are usually in talik, unlike the surrounding areas. This creates potential problems. In the case of the larger rivers, the pipeline was slung on bridge supports anchored in the talik, e.g. the Tanana river crossing, which is 360 m long. Either way, problems can occur with changes in the course of the river. This sometimes means that the river pattern needs to be controlled or modified, but this must be done in such a way that the conditions necessary for the survival of fish are preserved.

No reliable method has been developed to prevent icings. Freezing of the lower portions of both the service road and the pipeline causes icings in certain areas by interfering with the local drainage. Sometimes steam jets are used to thaw frozen culverts but they are expensive and inefficient. Better drains and drainage have helped but the location of near-surface drainage varies from year to year. The worst problems arise when the ice engulfs the VSMs, whereupon expensive and difficult removal is essential to allow the pipe to move in response to temperature changes or earthquake tremors.

The gas and oil storage facilities near Prudhoe Bay and elsewhere on permafrost were placed on piles elevated 2-2.5 m above the ground surface. Gravel pads were used under small buildings, while refrigeration systems were built into the pads for many of the pumping stations where power was no problem.

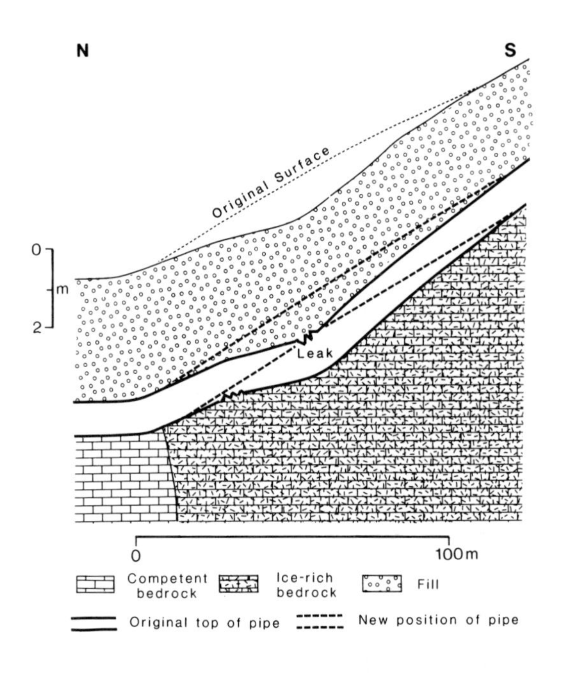

Figure 7.17: Conditions at the time of Failure of the Pipeline at Mile Post 166 on the North Slope of Atigun Pass, Brooks Range, Alaska (after E. Johnson, 1981).

Failures in the Buried Section

E. Johnson (1981), Thomas et al. (1982) and Stanley and Cronin (1983) have described the situation at two sites where failure of the buried pipeline has occurred. One was at Mile Post 166 on the north slope of the Atigun Pass in the Brooks Range, while the other was at Mile Post 734, about 800 m north of Pump Station #12, about 151 km north of Valdez.

At Mile Post 166, drillholes had indicated unstable ground in many places but avalanches and rockfalls in the mountains caused it to be desirable to bury the pipeline and use the "insulated box" technique with 50 cm of high density polystyrene insulation but no refrigeration.

Preconstruction drillholes near the site indicated the presence of ice in the bedrock profile. Although some ice was observed during trenching in the walls of the ditch, the underlying rock appeared

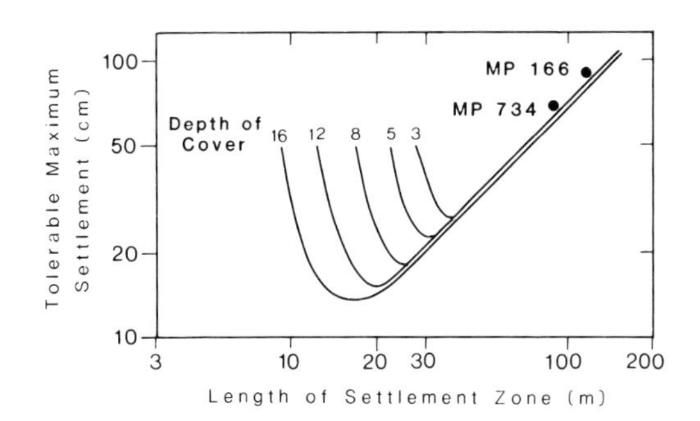

Figure 7.18: Maximum Subsidence Designed for on the Alyeska Pipeline Compared with Conditions at the Time of Failure at Mile Posts 166 and 734 (after E. Johnson, 1981).

ice-free. Accordingly, conventional burial was used.

Failure occurred in June 1979 after twenty-four months of operation. Sixty boreholes were drilled around the leak and the surface of both the ground and the pipe were relevelled during excavation(Figure 7.17). Thaw settlement of up to 1.22 m had occurred and was reflected in subsidence of both the ground surface and the pipe along a 122 m length. It was also discovered that this section had been laid on bedrock with ice masses up to 3.3 m thick in it and having an overall visible ice content locally reaching 50-90% by volume. This occurred at depths from 3-7 m below the surface but ice was absent below about 31.8 m. Maximum subsidence designed for was 0.91 m (Figure 7.18), so the pipe had performed as expected. Leakage occurred in the upper of the two buckles with a bending of 3°.

Subsequent drilling in 1979 in a similar area on the south side of the Pass revealed excessive thawing, thaw settlement and ground water flow. Surface water at the crest of the Pass was disappearing underground, much of it into the trench containing the pipe. Provision of better surface drainage and grouting, followed by mechanical refreezing solved the problem (Stanley & Cronin, 1983).

The leak at Mile Post 734 occurred about one week later but at the southern limit of the sporadic permafrost zone. It is in this zone that prediction of the permafrost distribution is most difficult. Although the ditch logger recorded

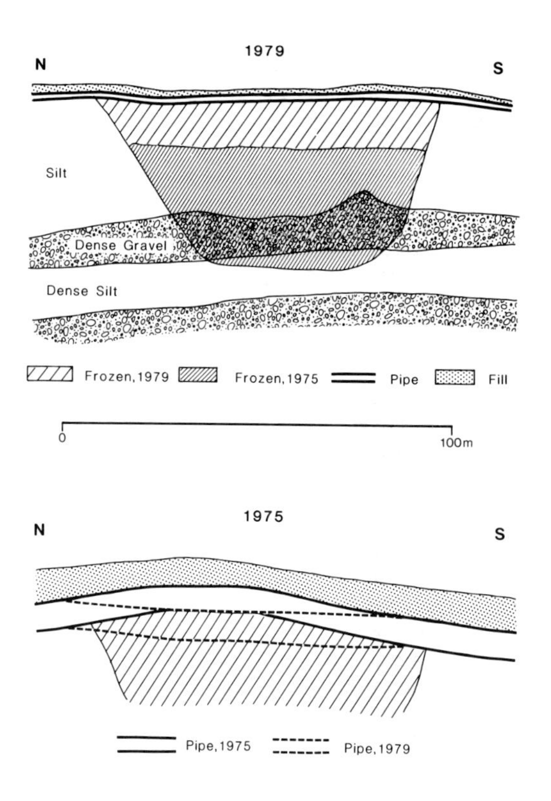

Figure 7.19: Conditions at Mile Post 734 at the Time of Failure (after E. Johnson, 1981).

"thawed" material in the trench bottom in May 1975, drillholes after the leak showed that frozen, highly unstable permafrost was present along a 91 m length of the pipeline (Fig. 7.19). Up to 75% visible ice was present in the surficial material around the break. Over 2 m of settlement had occurred under conditions where only 62.5 cm of settlement had been designed for: once again, there was a 3° bend that caused rupture.

The alternative methods of repair were rerouting the pipeline, subsurface grouting, mechanical refrigeration, or underpinning of the replacement pipe with piles embedded in the deeper stable zone. In both cases, the last alternative was chosen as being the most convenient and economical.

A study of the pipeline where failure had taken place showed that thermokarst features developed prior to the failure (Thomas & Ferrell, 1983). Thus a thaw lake developed at mile 735 and study of the surface of the buried sections of the pipeline can indicate clearly where substantial thaw settlement is occurring.

In early 1985, the pipeline had to be shut down while a replacement loop pipe was constructed for the crossing of the Dietrich river. Thaw subsidence of several metres had occurred but was only detected when in an advanced state in spite of the monitoring programme. Fortunately, no great environmental damage occurred.

Monitoring

Since the failures in 1979, both the above-ground and buried sections of pipeline have been monitored on a regular basis, as this is cheaper than repairing failures. Settlement rods have been attached to the buried pipe at 1,100 locations and at least 450 additional soil borings were made between 1977 and 1981 to check out soil conditions during operation. Four hundred thermistor strings had been installed to check the thermal conditions around critical sections of the pipeline. Third order survey benchmarks have been installed at 1.5 km intervals along the route to check for settlement of the buried pipeline or heave of the VSMs. These allow for identification of as little as 6 mm vertical movement. At least 12 heave rods are tied in to each benchmark, thus giving comparative readings as a further check. By 1981, only about 550 VSMs had heaved. E. Johnson (1984) describes the performance of the pipeline.

Monitoring the pipeline is minimised where the pipe is found to be stable and is concentrated at known locations of potential settlement. Appropriate repair plans and contingency measures can then be developed before a leak actually occurs. Data from all the monitoring sites are collected every month and stored and checked by computer. As noted earlier, the heat exchangers on the VSMs are also checked, using infrared photography, so that any that prove faulty can be replaced.

By carrying out this monitoring, further failures have been detected before a major oil spill occurred. It has been found that this monitoring programme is much easier and less costly than oil spills. Even though the above-ground mode leaves the pipe open to sabotage and is aesthetically less satisfactory, it is easier to repair than the below-ground sections.

GAS PIPELINES

It is safer to have gas pipelines in the buried mode if possible. This reduces the range of temperatures to those where steel is sufficiently tough to withstand the pressures and it also reduces the risk of explosions. With small diameter pipes and/or lower pressures, the pipe can withstand greater stresses without failing. As the diameter and pressure increase, the quality of the steel and the thickness of the wall of the pipe must be increased. Good welded joints become critical since these tend to be points of potential weakness. In addition, external stresses due to thermal expansion and contraction or heaving and thaw subsidence must be reduced to stay within the acceptable limits required by the design.

It will be seen in Figures 7.13 and 7.14 that there are usually only small collector pipelines for gas in permafrost areas, although multiple large diameter pipes are common elsewhere. This reflects the difficulties of construction rather than lack of gas in permafrost areas.

The first gas pipeline constructed over permafrost appears to have been between Tas Tumus and Yakutsk shortly before 1970. The pipeline is 52.9 cm and 32.5 cm in diameter and extends for about 600 km. Sixty percent was built elevated on piles, 8% was laid on the bare ground and the remainder was buried. Choice of mode of construction was based on the results of drillholes, as in the case of the Alyeska oil pipeline, which was constructed later. Pressure in the pipe was 3.5 Mpa. Another small diameter gas pipeline was buried alongside the Alyeska pipeline to supply the pumping stations on the north slope of the Brooks Range, Alaska, while a third system was constructed to supply Norilsk with gas in 1968-9. It is entirely above ground except for river crossings (Ferrians, 1984).

In spite of this, in 1971 the Director of the Permafrost Institute of the Soviet Academy of Sciences stated that large diameter gas pipelines in permafrost would present greater problems than the small diameter pipelines constructed until then, and had not been studied (Slipchenko, 1972). During the 1970s, large diameter (1.2 m or larger) pipelines with gas pressures greater than 7 MPa were suggested and designed. However, so far, none have been constructed.

The major problem encountered is frost heaving and its control. Due to the extremely high pressures and the large pipe diameter, it is undesirable to leave the pipe above ground. However, with burial, a chilled gas pipeline will tend to cause ice segregation and heaving in the surrounding soils. There have been numerous studies of the problem and, at first, it was thought that heaving could be controlled by piling sufficient earth on top of the pipe, so that its weight equalled or exceeded the heaving pressure. Unfortunately, the experiments supporting this proposal were based on using a pressure apparatus that was leaking, and it has since been found that the height of the berm required would need to be a minimum of 30 m (P. J. Williams, 1979). The Ad Hoc Study Group on Ice Segregation and Frost Heaving, Committee on Permafrost of the United States Polar Research Board has presented an excellent discussion of the literature and concludes that continued research is essential (Anderson et al., 1984).

Use of higher temperatures for the gas will result in thawing of the permafrost around the pipe and resultant thaw subsidence. This can be reduced considerably by insulation of the pipe (Jahns et al., 1973) but it cannot be prevented.

Until these problems can be overcome in an economically acceptable way, building large diameter pipelines remains impractical.

Plate XVII: Stripping Silt at Ready Bullion Creek, Near Fairbanks, Alaska, in 1965.

Plate XVIII: Gold Dredge Used for Working Gravels, Ester Placer Mining Area, Cripple Creek, Alaska, in July, 1965.

Chapter Eight

MINING

Mineral resources are abundant throughout the Arctic, resulting in a major mining industry. Originally, the mines were chiefly located close to markets and transportation routes in the non-permaforst areas such as the Ural mountains, south-eastern Alaska, and southern Ontario, but within the last century, very productive mines have been established throughout the permafrost regions (see Fig. 8-1). To do this, it has been necessary to modify normal mining practices considerably.

Three main problems occur when mining in permafrost, viz.: the coldness of the deposits, the ice content, and a restricted season for transportation. Methodology is steadily improving but mining in permafrost is inevitably much more expensive than in other areas. As a result, the ore has to be very high grade, such as the asbestos at Cassiar (British Columbia), or contain very valuable minerals, such as gold at Dawson City (Yukon) or diamonds at Mirny (Siberia), or it has to be needed locally, such as the coal at Svalbard (Spitsbergen). Proximity to low cost transportation, e.g., on Baffin Island, or to major markets, e.g., Schefferville, can cause exceptions to these rules.

For convenience, the mining projects can be divided into placer, open-cast (or open-pit) and underground workings. Each group has a different set of problems.

PLACER MINING

Martin Frobisher encountered ground ice during his abortive attempts at gold mining on Baffin Island in the 1570s. Later miners had similar problems with permafrost in North America in the gold rushes in the Yukon Territory and Alaska. The gold

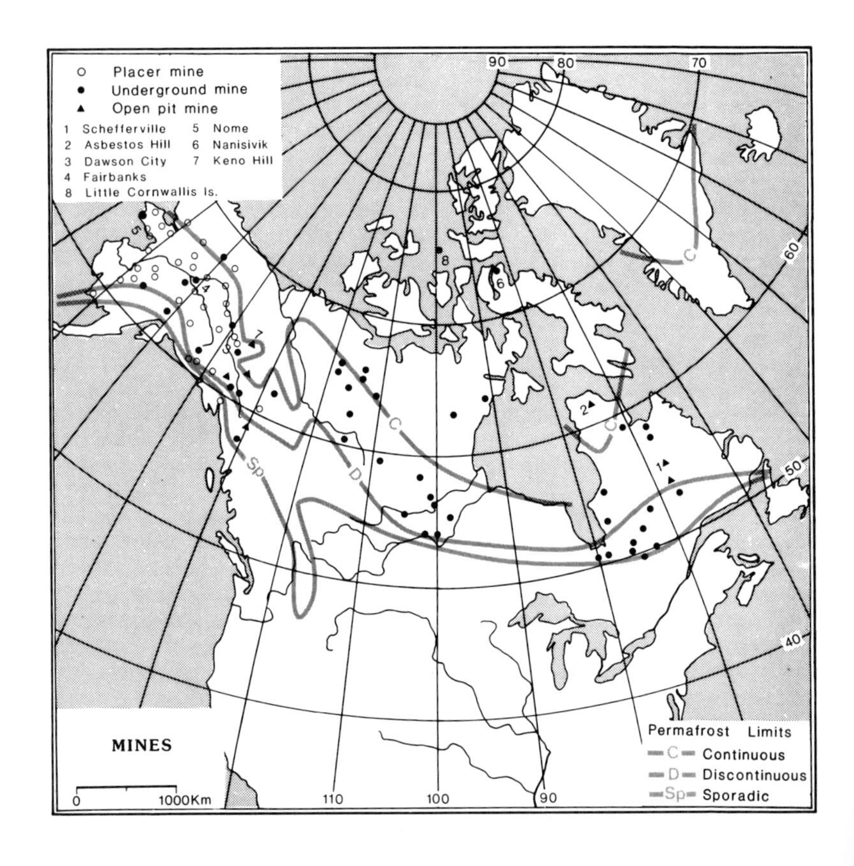

Figure 8.1: Locations of Mines in the Permafrost Areas of Canada and Alaska after Geological Survey of Canada (1983) and Alaska Geographic (1982).

occurs in gravel placer deposits below a thick overburden of silty deposits such as loess and peat. The overburden is locally called "muck" and may be up to 30 m thick.

The gold-bearing gravels probably represent outwash deposits from streams fed by expanded glaciers in the nearby mountains. Near Dawson City, at least three different Pleistocene gravels are present (Naldrett, 1982), each probably formed after substantial nonglacial periods of at least 300,000 years (T. L. Péwé, 1983, personal communication). The individual gravel deposits contain 10-12% ice by weight and may be up to 30 m thick. The oldest gravels (the White Channel gravels) occur on benches

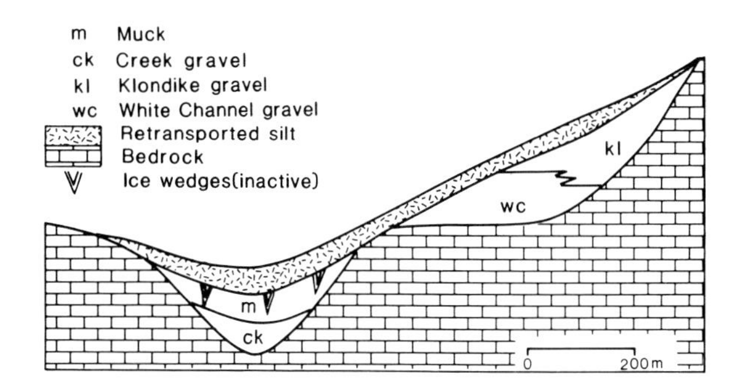

Figure 8.2: Typical Distribution of Valley, Creek and Terrace Gravels in the Klondike Area. Terrace gravels occupy upper benches, while valley and creek gravels occupy the lower smaller valleys (from French et al., 1983).

high on the valley sides and are completely frozen. They are overlain by a second, younger gravel deposit (Klondike gravels, Fig. 8.2). The youngest gravels (Creek gravels) on the valley floor are overlain by about 30 m of "mud" containing up to 44% ice by weight as well as substantial ice wedges of two different ages (French et al., 1983). The present-day active layer is only 50-70 cm thick but the tops of the ice wedges lie at 150-170 cm depth. This is probably due to deeper thawing during a warmer interval such as the Hypsithermal event. An excellent collection of photographs showing the conditions during the Gold Rush will be found in Bolotin (1980).

The first gold rush was to Dawson City in 1896. The original mining was carried out by sinking vertical shafts through the frozen silts into the gravels. The latter were brought to the surface for working but this involved a tremendous effort for limited rewards.

Then the miners used the heat of the sun to thaw the surface layers during summer and scraped off successive thawed increments until the gravels were exposed. Then the gold-bearing gravels were thawed and worked through sluices. However, it was terribly slow and could only be used in summer. Artificially thawing the ground by burning wood fires on the surface or dropping hot rocks into the workings permitted year-round mining, although sluicing still had to be carried out in summer.

By 1909, when the gold rush ended, the area around Dawson City was completely deforested.

In 1900, hot water and steam were tried but resulted in many accidents. By 1899, placer deposits were thawed out by various means including thaw points, hot water points and, later, cold water points (pipes) were used extensively. In 1917, the technique of hydraulicking was introduced from California, and still continues to be the main basic method for thawing the deposits in Canada, but is banned in Alaska. It consists of spraying jets of water under high pressure from large monitors fed by pumps or gravity on the frozen deposit and is relatively fast and cheap. A special canal and pipe were used to bring water under pressure into the area from the nearby hills. Giant dredges were brought in and assembled. They were powered by electricity from hydroelectric plants and worked the thawed gravels which had accumulated in giant settling ponds on the valley floor. Remnants of the resultant tailings that form parallel piles of gravel can be seen on most of the valley bottoms east of Dawson City. The remains of one of the last of the dredges is to be found in the National Park in Bonanza Creek, although two working dredges could still be seen in 1981 at Nome, Alaska. Similar methods were used at Fairbanks, Alaska (Plates XVII and XVIII).

Modern workings involve the use of bulldozers, sluices and monitors, i.e., spray guns for the water (French et al., 1983). The sediments on the valley sides have also proven to be auriferous and are now being worked. During the original gold rush, the gravels on the high terraces could not be worked but since the price of gold increased dramatically in 1974, it has become an economic proposition to pump water to the top of the slope and wash the more accessible gravels down to the sluices. If the price of gold remains high, heavy earth-moving equipment will be used to move the gravels to the sluices.

Determination of gold content is achieved by augering holes, saving the material for assay and then determining the volume of the displaced material by filling the hole with a measured volume of cold water (Beistline, 1963). The permafrost prevents loss of water. The gold content is determined on augered material.

To work a claim, a water supply must be secured and the vegetation stripped away. The unfrozen overburden is removed by bulldozer and the silts are thawed and washed away with the monitor. This

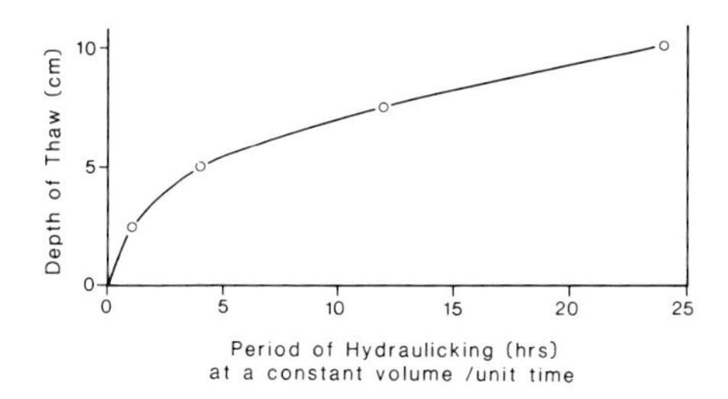

Figure 8.3: Typical Thaw Rates of Silt at Fairbanks (data from Johnston, 1948).

creates considerable silt pollution in the streams which renders this section of the Klondike river and its tributaries useless for fishing. This is not permitted in newer workings elsewhere in Canada, where settling ponds are required to protect the aquatic life in the streams. The underlying gravels are thawed and then pushed by bulldozers into piles ready for passage through the sluice box.

The rate of thawing of the ground by the water methods depends on the ice content, the grain size, the thermal properties of the material, the temperature of the ground and the temperature of the water. Organic deposits thaw faster than gravels, while silt takes the longest time to thaw. Figure 8.3 shows typical thaw rates for frozen silt at Fairbanks, Alaska. About 80% of the heat is used in melting the ice, so that the ice content of the frozen material is rather crucial in determining the rate of thaw.

The water may be applied to a slope as a spray or it may be applied as a sheet to a flat or gently sloping surface. However, this is only usable on floodplains and low terraces, which are uncommon in Yukon and Alaska. In Russia, the water is often spread over the prospect to a depth of 5 m (Brown, 1970a) and is left to thaw the ground for several years. Provided the water does not freeze to the bottom in winter, this results in thawing of the underlying deposit, which can then be dredged. In Alaska, the Miles method applies water at about 10°C in a grid pattern through steel pipes which are pushed into the flat surfaces of the ground and the pipes can be pushed downwards as thawing proceeds (Beistline, 1963). The Pearce method applies water evenly to a sloping surface, allowing the water and thawed material to run off to a settling pond. Then the water can be recycled. However, this method went out of use because of its relative ineffectiveness. At Nome, Alaska, cold water was applied through pipes driven down to the bottom of the auriferous gravels and was allowed to percolate upwards (Kastelic, 1982). On the way, it warmed the gravels so that they would be thawed and ready for dredging after two years (D. J. Cook, 1983).

The Soviet Union has a different economic and political structure from North America, which results in the use of four thawing methods which cannot normally be duplicated on the other continent. The successful use of slow annual removal of the thawed active layer from extensive prospects each year has been reported from the northeast of the USSR (Emelyanov & Perl'shtein, 1980; Pavlov & Olovin, 1974). Formulae have been developed to predict the probable depth of thaw and ground surface temperature according to July air temperatures. Thickness of thawed material produced each year ranges from less than 30 cm in the north to a maximum of about 120 cm in the south. Actual thickness of material removed depends on its lithology and, especially, its ice content. This produces a considerable local variability, i.e. 20-50 cm for a given climatic situation.

A second method is to speed up this thawing by using hot industrial waste water either directly, or through heat exchange systems to thaw the ground. This includes waste water from atomic power stations. The first of these low temperature heaters was developed in 1974 in Magadan Oblast. This has now been developed to a point where the water can be applied to completely frozen ground and heat losses to the atmosphere kept at a minimum.

The third method involves the use of tubular needle electrodes for thawing large masses of frozen ground with currents of electricity of industrial frequency. The electrodes are placed so as to form a network of equilateral triangles; each electrode is equidistant from six adjacent electrodes and interacts with them. Perl'shtein and Savenko (1977) have developed an approximate method for assessing the heat produced and it is apparent that expansion of the thawed zone is accompanied by increasing unproductive heat losses to the atmosphere. Thus, the electricity supply needs to be intermittent, or else the applied voltage needs to be reduced gradually with time to economise on power. Obviously, this system can only be economic where there is excessive cheap electrical power available.

The fourth method, which is just being developed, is the use of hot geothermal water from deep in the ground. It remains to be seen how well this works since the cold permafrost zone can be very thick and the hot water wells would need to be very deep. Experience in North America suggests that problems may be encountered with corrosion and blocking of the pipes by the mineral-laden waters as the water cools.

The Russians are also using a thawing-drainage technique for producing "sushentsy", i.e., loose deposits with too little moisture to be strongly ice-bonded after freezing. The method involves cutting numerous closely spaced, shallow drainage-ways to remove the water from the deposit as it thaws. This greatly speeds up thawing and drainage and produces a gravel deposit which can be moved by earth-moving equipment even at low air temperatures during the autumn and winter (Emel'yanov, 1973). Ideal moisture contents for this are under 3% by weight (Emel'yanov & Perl'shtein, 1980). This filtration-drainage thawing method remains a slow procedure for working the deposits. Very extensive prospects are needed in order for it to be an economic proposition in North America.

Yet another modification is the circulation of water in polygonal cells, moving downwards through the permafrost to a central drainage cell. The water is pumped out of the cell and recirculated, being warmed by the air and warmer ground surface on the way (Chaban & Gol'dtman, 1978).

Because of the high cost of working placer deposits in permafrost regions, the only minerals for which it is presently economic are gold, the platinum group minerals, tin and tungsten (Bundtzen, 1982). In North America, these are primarily found in the outwash deposits from the Pleistocene mountain glaciers in the northwestern portion of the Cordillera (see Figure 8.1).

OPEN-CAST MINING

The techniques used in the mining of placer gold deposits in permafrost regions cannot be applied to open-pit mining because of the deeper, coherent ore body, often situated on a hill or mountain, and having a limited areal extent. Drainage and surface water must be removed continuously to facilitate safe working of the

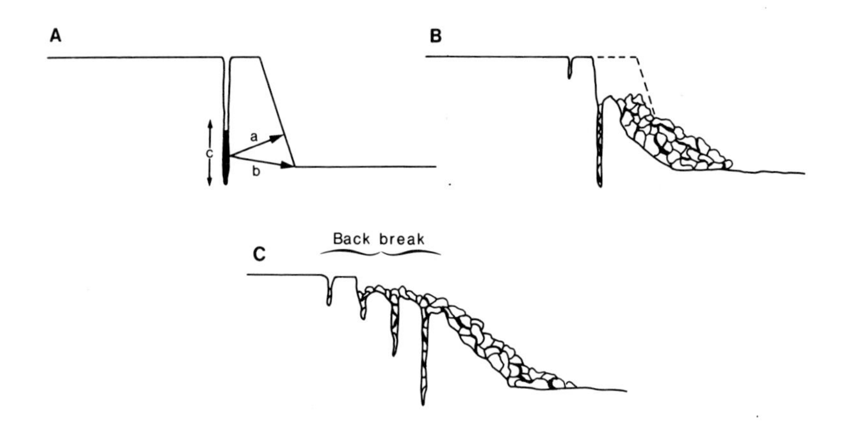

Figure 8.4: (A) Idealised Mining Face, (B) Reasonably Successful Blast, (C) Poor Blast in Open-cast Mining (modified from Ives, 1962). In drawing (A), (a) is the critical distance between the centre of the charge (c) and the mining face (called the "burden"), while (b) is the distance between the toe of the slope and the centre of the charge (c) (called the "toe distance"). If the toe distance greatly exceeds (a), extensive secondary drilling and blasting are necessary to correct the slope of the face.

deposit. Slope stability is most important. The ore and overburden are usually frozen and may contain variable but usually significant quantities of ice. Solar thawing is impractical. The lower porosity of the rock usually results in lower overall ice content than in detrital sediments.

The usual way of freeing the ore is by drilling and blasting the rock so as to produce a fractured mass of aggregates of a suitable size for transport and processing. Too small a charge produces fragments that are too big to place in a dump truck, while too large a charge will pulverize the ore. The blasting (Fig. 8.4A) should free a convenient mass that can be removed before it refreezes. The base of the shattered area should be even and flat and the back-face nearly vertical (Fig. 8.4B). If the base is sloping, then the loose material is more difficult to pick up with the power shovel, while secondary blasting of the sloping bottom of the pit will be necessary to produce a suitable back-face for the next major blast (Fig. 8.4C).

Considerable experimentation has taken place in an attempt to determine the best way to blast frozen ground in open pits. Comprehensive summaries will be found in USSR (1972), Morgenstern et al. (1978), and Bauer et al. (1965; 1973).

The best documented case of improvement of opencast blasting techniques is that of the Iron Ore Company of Canada mines at Schefferville, Quebec. Between 1970 and 1982, a major research programme was carried out to improve the techniques. The mine lies in the discontinuous permafrost zone and ground temperatures varied from 4°C to -2.5°C. Ice content varied considerably within the frozen ground.

Mining began in 1956 and the first problems encountered were with drilling holes for blasting. If water was used, then the drill string and rod would often freeze in down the hole, making them difficult to retrieve. Without a fluid , drilling tended to heat the walls of the hole and melt any ice present, and the resulting water could then refreeze, often freezing the drill string in the hole. Melting of the sides of the hole also caused slumping and collapse of the walls if the hole was left open too long. This resulted in the improper placement of charges or the need to redrill the hole, with accompanying increases in production costs.

These problems could be partly overcome by drilling with air and using suction to retrieve the cuttings. Calcium chloride solution also allowed wet drilling at temperatures below zero but did not prevent thawing of ice in the walls. It corroded metals including drilling bits, it cost more, and extra time was taken to mix the solutions.

The drilling fluid was sucked out of the hole if the hole was less than 12 m deep, so the explosive could be packed in the dry hole immediately after drilling. Although the cold ground facilitated "stemming", i.e., sealing the charge in the ground, melting and slumping of the ice in the wall often prevented proper placement and detonation of the explosive charge. This was overcome by using a plastic liner for the hole before the charge was inserted. Dry sand or rock chips were used for stemming and the usual cheap explosive employed was ANFO (a mixture of ammonium nitrate and fuel oil).

Ice is elastic and it requires a larger charge to break up ice-cemented rock containing excess ice. However, gravels or rock with only pore ice can be blasted in the normal way. Indeed, the ice increases the shear strength of the rock so that less charge is needed than in non-frozen rock.

The early work on blasting rock with excess ice followed the procedure of Livingstone (1956), varying charge size, depth of placement and spacing to obtain the optimum maximum crater volume for a given number of charges.

Bauer et al. (1965) provide typical results of tests with placement of explosive charges in seasonally frozen ground in different materials.

Even with the increased knowledge provided by various studies, the Schefferville iron mine still encountered problems due to the high variability in the temperature and the ice content of the ore. These conditions produced an uneven base to the bottom of the pit which made the use of mechanical shovels for loading the blasted material very difficult. Material that was broken up and not removed quickly refroze during much of the year, creating additional problems.

Accordingly, a further study was made to determine the most efficient way in which both ground temperature and ice content could be accurately determined prior to working a face (Garg, 1977). Experiments showed that the geophysical properties of the ore varied with ice content and that unfrozen zones could readily be distinguished (Garg, 1973, 1979; King & Garg, 1980). Tests proved that seasonal frost with ice lenses, unfrozen zones (taliks) and zones of continuous permafrost were present in the ore body. The taliks were usually associated with excessive winter snow accumulation or small water bodies in depressions (Nicholson, 1978b, 1979). A system of regional exploration and deposit evaluation was devised to overcome the problem by producing a three-dimensional picture of the distribution of the ore, the overburden and the permafrost.

Experimental blasting was used to establish the correct pattern and depth of the charges for a given situation. Time lapse photography showed the order in which the charges were detonated to produce the most efficient fragmentation of the ore. Thereafter, charge placement was designed using a computer.

Only a ledge wide enough to be worked by a given shift was blasted. This avoided refreezing of the ore. The charge patterns developed by Garg (1982) are shown in Figures 8.5 and 8.6. In ice-cemented ground containing excess ice, the spacing is much closer and two types of drillholes and charges are used. One is designed to cope with the seasonally frozen ground and the other with the

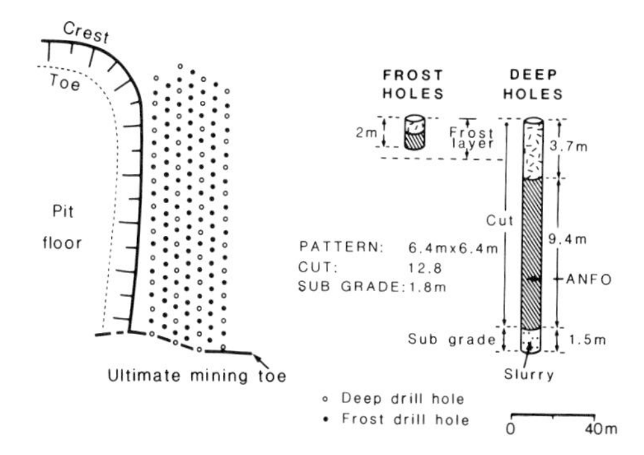

Figure 8.5: Typical Blast Pattern in Ice-cemented Ground Containing Excess Ice (Garg, 1982, p. 590). Reproduced with the permission of the National Research Council of Canada.

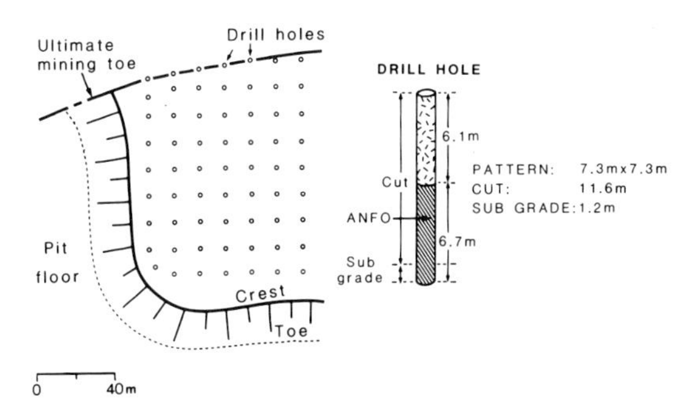

Figure 8.6: Typical Blast Pattern in Unfrozen Ground or in Ground Containing Pore Ice (Garg, 1982, p. 589). Reproduced with the permission of the National Research Council of Canada.

permafrost. In unfrozen ground, only the deeper charges are used, the spacing being considerably wider and the size of the charge being only about 70% of that used in frozen ground. The bulk of the charge is now aluminised ammonium nitrate and fuel oil mixture (Al ANFO) which can be inserted down the collar and plastic tube into the hole. Special precautions still have to be taken where the ice content is high but the programme resulted in substantial savings of money and considerably greater efficiency.

Unfortunately, when the demand and prices for iron ore dropped during the 1980 recession, mining was halted; the cost of mining and transporting relatively low grade ore was too great. Nevertheless, the experience gained at Schefferville can readily be applied in modified form elsewhere.

Successful open-pit mining is more complex in areas of discontinuous permafrost due to the variable ground temperatures on either side of the freezing point in the ore body. Ice content can be expected to show considerable variability in all frozen ore bodies and correct determination of the charge spacing and depth can only be made after a thorough three-dimensional study of the ice content and temperature of the ore body. In very cold regions, the need to move fragmented material immediately to avoid refreezing is more acute.

Successful open-pit mining is carried on for a variety of ores. The low grade blue ground from the diamond mines at Mirny, Siberia has to be processed close to the mine in order to keep down costs but the diamond production from this mine is sufficient to supply the entire needs for industrial diamonds for the USSR and the Eastern European countries. The open-pit method is usually the cheapest of the types of mining because of the smaller operational cost per tonne of ore; hence it is more economic than underground workings for mining lower grade ores.

UNDERGROUND WORKINGS

Many high-grade ore bodies are highly inclined, relatively deep, thin, clearly delineated or discontinuous. Underground mining is the only practical way of working these deposits economically.

Fernette (1982) describes the main ways of mining underground. Key factors include the geometry, strengths and stability of the host rock and the ore body. Permafrost affects these strengths and stabilities, the actual modification varying according to the ground temperature, the temperature of the air in the workings and the ice content of the rock. Most deep mines in permafrost areas in North America use ventilation by air at 5-10°C. This will cause thawing of the ice in the walls of the rock; the actual amount of thaw will vary depending on the time of circulation of the warm air, the temperature of the bedrock and its ice content. Typical predicted results for various depths at Asbestos Hill, Quebec are shown in Figure 8.7 (after

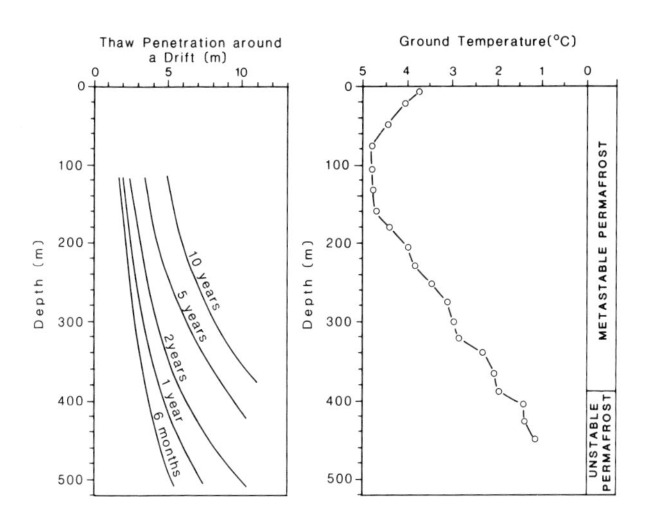

Figure 8.7: Predicted Extent of Thaw Around a 3 m Diameter Drift or Decline at Various Depths Compared with the Ground Temperature and Permafrost Stability at Asbestos Hill, Quebec (modified from Taylor & Judge, 1980). Thawing is based on 5°C air circulated continuously for time periods between 6 months and ten years. Note the greater thaw in the warmer ground at depth in a given time. Reproduced with the permission of the National Research Council of Canada.

Taylor & Judge, 1980). However, if the mine is kept cold by circulating cold air, the host rock and ore body will be stronger than in the unfrozen state. This is the practice usually followed in the USSR and by Cominco on Little Cornwallis Island.

A mine typically is designed with a main entrance for taking in men and supplies and removing the ore. In addition, the mining operation will open other holes called "stopes" as the ore is removed. The actual mining is accomplished by blasting at the end of each shift when everyone is out of the mine. Then the loosened ore is removed and hauled to the surface by truck, train or conveyor. This is followed by drilling and emplacement of the next charges.

When stopes or raises break through to the surface or thawing of the permafrost takes place, glacier-like ice masses may develop in the openings due to the melt-water refreezing as it moves down the passage through the permafrost. These ice

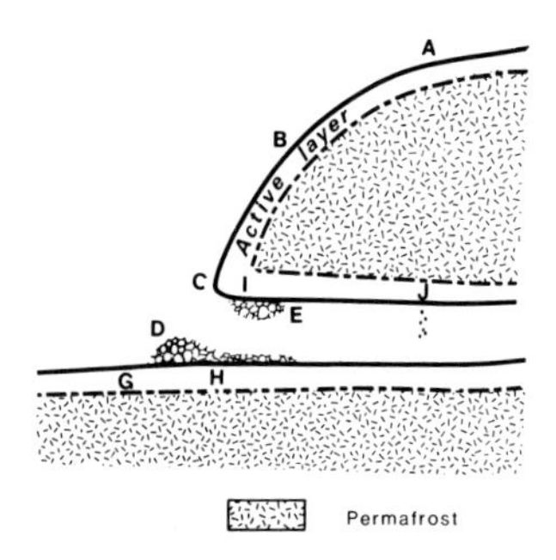

Figure 8.8: Problems that may Occur Around Tunnel Entrances (modified from Linell & Lobacz, 1978, p. 814). Reproduced with the permission of the National Research Council of Canada.

masses in the tunnel are virtually impossible to stop.

The mine entrance is a common source of trouble. If a headframe is used at the top of a vertical shaft, any differential heaving or settling caused by ice segregation or thawing will cause problems. Eldorado Mining Co. removed a peat layer and built a headframe on 10 m of frozen ice-rich silt. Very soon, a new thaw lake formed and the shaft and headframe had to be abandoned (Brown, 1970a). This can be overcome by using an internal headframe and shaft.

Where possible, an entrance in the side of a hill is preferred. Even then, many problems may be encountered (Fig. 8.8, after Linell & Lobacz, 1978). The problems shown at the lettered locations are:

- A Surface disturbances that may cause degradation of permafrost.
- B Uncontrolled drainage that may cause erosion, icings, and accelerated permafrost degradation.
- C Slope creep or closure of the tunnel that may decrease the height of the tunnel and damage the portal structure.
- D Falling slope material which is a hazard to people, while slides or sloughing can block the entrance.
- E Material that may fall from the roof with thawing.
- F Air temperature in the tunnel that may drop below freezing.
- G Uncontrolled drainage water that may flow into the tunnel.
- H Portal structure that may locally increase the depth of thaw.

I Seepage entering from annual thaw ones that may form hanging roof ice masses and water and that may collect and freeze on the floor of the tunnel.
J Thawing of ground ice that can cause water seepage.

Melting of ice masses will cause water to enter the workings from the walls, along with water contained in any aquifers or taliks that may be intersected by the workings. Flooding of mines can be disastrous and occasional tragedies occur when meltwater suddenly enters the mine and pours downslope.

Sometimes the portal area may need to be refrigerated to avoid excessive thawing and risk of collapse. Much of the heat brought into the mine by ventilation escapes through the portal. To control the temperature of the mine and to prevent cold air entering, double doors and air locks may be used. Timber and concrete are necessary to stabilize the ground, e.g., United Keno Hill mine, Yukon Territory. In general, it is easier to mine in continuous and stable permafrost than in unstable or metastable discontinuous permafrost.

Any water entering the mine must be removed immediately and continuously by a high pressure pump. If warm air is introduced, it will form icings on the surface of the colder parts of the mine, e.g., on the walls and on any equipment there. This ice must be removed regularly. The moisture can also interfere with power and communication cables and equipment. Air lines must have some methyl hydrate blown through them to prevent blockage by ice. If a pump removing water is shut down, it and all pipelines must be blown clear of all water. Sump-pumps must work continually to remove all water before it freezes.

Shut-downs of mines occur in North America whenever the price of metals falls too low. This entails draining of all air and water lines.

Immediately after blasting, the mine must be ventilated to remove the toxic gases and fumes. In mines with cold air ventilation, ANFO can be used as the explosive but where warm air is used, other more reliable but more expensive explosives may be required. Normally, two ventilation lines are run into warm air ventilated mines (Swinzow, 1963). At least one must follow the lowest part of the mine to move the colder air out of the way. Water is used to wash down the walls to keep dust to a minimum. This water must be removed efficiently and immediately by sump pumps. There is a tendency to produce spalling off of material from the walls if they are heated too quickly (Pettibone, 1973; Thompson & Sayles, 1972). In mines ventilated with cold air, the water may be left on the walls as an icy coating to add strength and seal in the dust. About 70% of the energy in the cold air may be used in this way (Emel'yanov & Perl'shtein, 1980).

In cold mines, there is reduced icing up of equipment compared with mines ventilated with warm air. However, all loosened ore must be removed quickly before it refreezes. The ore occupies 50% more volume after it is blasted. Sublimation of ice from frozen silts and gravels results in the production of dust on floors and in the ventilation systems.

Strength of the rock is greatly modified by the ice content and the temperature. The air used in ventilation has to be humidified after warming to avoid sublimation of ice and loss in strength of the roof and walls. Walls with temperatures below -4°C have much greater strength than those at -2°C to 0°C. The latter require many times more support in the form of props and the unsupported spans must be very short (Chaban & Gol'dtman, 1978). Where the mine is warm, suitable props must be emplaced immediately to allow for the anticipated thawing, loss of strength and spalling of the walls. It is both dangerous and expensive to add extra supports later.

When a stope or adit is mined out, mine waste from other production faces or non-toxic mill tailings may be brought in and backfilled into the space and allowed to refreeze. This reduces the need for long-term supports to maintain mine stability. In cold ventilated mines, this refreezing is very rapid. Backfilling rooms with pillars will allow the remaining pillars of ore to be mined, which can increase the recovery from 60% to 90% of the potential tonnage. Infilling of the surface waste can only be carried out in summer when such material is sufficiently thawed to be moved.

Electrical machinery is preferred to diesel due to the fumes of the latter, which require far greater ventilation. The diesel exhausts also add a significant amount of heat that must be allowed for.

SUPPORT FACILITIES

Men and supplies have to be brought in, bunkhouses and warehouses constructed, and power supplies provided. Water supply has to be available for camp and industrial use throughout the year and tailings ponds and waste heaps constructed. All these have to be built in such a manner that they will not suffer any damage that will affect their use (see Chapters 4 and 9).

TOXIC WASTES

These can be stored in old mine shafts and workings, provided enough is known about the geology to ensure that the wastes cannot escape through the rock or in the circulation of groundwater, e.g., arsenic at Yellowknife. They must be sealed off by specially designed bulkheads to ensure that they cannot escape, even if the mine eventually becomes flooded (Lopez et al., 1984). Regular monitoring is essential to ensure that any environmental damage can be identified and dealt with.

TRANSPORT OF THE ORE AND WASTE

Dry, frozen materials can be handled at any time of the year without problems. Hard, frozen material is easy to move as long as it is freshly broken up and kept cold. However, both materials produce dust, which causes problems with bearings and adheres to heated surfaces. Wet thawed or thawing ore is usually messy and freezes to any cold surface. Between 0°C and -10°C, these problems are most acute. The force required to break up frozen or refrozen material increases threefold between -4°C and -25°C.

There are two basic problems to be considered, viz.: transport of material around the mine and transport to market.

Transport of Frozen Material Around the Mine

This has been discussed by Aitken (1970) and Foster-Miller Associates (1965). There are three basic methods used, viz.: belt conveyors, trains and load-haul-dump vehicles. Slurry may be transported in pipelines. In modern mines, trucks and self-propelled scrapers are used wherever material needs to be moved distances less than 2 km. For distances less than 900 m, screw conveyors, vibratory conveyors, grucks, bulldozers, loaders and rail-cars can be used. Railless systems (trackless mining) overcome the maintenance problems of rails caused by icing, heaving and subsidence (Bakakin, 1978).

Belt conveyors can be used in mines but the whole system must be either above or below 0°C (Dubnie, 1972). Below 0°C, the materials must be dry to avoid icings and freezing of the ore to the conveyor belt. Pure rubber belts have to be used, since synthetic rubber becomes stiff and brittle. With suitable precautions, e.g., using low temperature lubricants, they will operate down to -40°C.

Shuttle cars are commonly used underground but removing the build-up of fine particles and ice on the rails and cars can be a problem. Salt or calcium chloride can be sprinkled on the cars but tends to corrode them and make the surface damp. The best protection is to keep the ore as dry as possible and periodically clean the equipment. Ice must be removed periodically from all stationary swithes. The advantages of keeping the mine frozen are obvious, and it has been found that with suitable clothing, men can work 9 hour shifts at temperatures down to -20°C, e.g., at the Polaris mine on Little Cornwallis Island. In northeastern USSR, a mine temperature of -5°C to -8°C is used since it approximates the ground temperature (see Chaban & Gol'dtman, 1978). This largely avoids the formation of fog and icing of machinery in the mine. The air is humidified to reach a maximum relative humidity of close to but not exceeding 85% to avoid desiccation of the mine. Hydraulic and pneumatic equipment are usually impractical unless special precautions are taken.

The waste and ore must be piled so that any thawing results in immediate drainage of the excess water. This will make it easier to move the material later. Unless the raw or processed ore is loaded or dehydrated or stored in a warm, dry environment immediately, it will tend to form a frozen mass. Piles of ore or waste can only readily be moved in summer when the surface thaws, although the development of giant rippers has helped extend the season somewhat. This material refreezes fairly quickly if taken back into the mine and used as fill, especially if the mine is kept frozen. In the USSR, water may be sprinkled on the fill to give it extra strength after emplacement.

Transport to Market

A basic decision is whether to process the ore to at least reduce the moisture content, to completely process the ore so as to only transport the end product, or whether to ship the ore. This usually depends on the ice content, value of the end product and amount of energy needed for processing. Thus, the ore from the diamond mine at Mirny, Siberia is processed at the mine site. A special dam for generating the necessary hydroelectricity was built about 100 km away at Chernosevsky. In that particular case, the value and small size of the product and the low grade ore established the most economical way of dealing with the problem. Similarly, gold is smelted at Yellowknife.

At Schefferville, Quebec, the iron ore plus ice was loaded onto railway cars and shipped to the coast of the St. Lawrence river for smelting. The ore was fairly pure, though rather ice-rich. The ore froze to the cars in winter so that shipping ore could only occur for 200 days a year. Later, the ice content was reduced so as to allow year-round shipping. Smelters with adequate energy supply were available within a reasonable distance of the coast.

At the Nanisivik lead-zinc mine at Arctic Bay, north Baffin Island, ore is processed until its moisture content is down to between 4 and 6% by weight (Northern Miner, 1974). It can then be handled without the fragments freezing together and without producing too much dust. It is then placed in ore ships and shipped south. To attempt to ship ore with higher moisture contents could result in severe handling problems during loading and unloading.

Plate XIX: Utilidors on Piles at Inuvik, Northwest Territories.

Plate XX: Road Passing Over Utilidor, Inuvik, Northwest Territories.

Chapter Nine

WATER SUPPLY, WASTE DISPOSAL AND PROVISION OF ELECTRICITY

Permafrost regions are characterised by extreme temperatures, low population densities and, usually, a poor economic base. They were often first settled by people who liked solitude and whose primary concern was to minimise the impact of cold winters. As a result, settlements are commonly low-density. sprawling entities, which considerably increases the costs of supplying essential services unless one resorts to resettlement. However, in North America this is only possible if the local inhabitants are in favour of it. Major factors which impede the provision of municipal services include permafrost, climate, remoteness, lack of planning for services, inadequate housing and lack of an adequate economic base to pay the costs of installing and maintaining the services (Gamble & Janssen, 1974; Smith & Heinke, 1981).

In the following account, the municipal services are dealt with under three headings, viz.: water supply, waste disposal, and provision of electricity. With the coming of telecommunications satellites, communication with the outside world and receiving radio and television signals are no longer major problems.

Guidelines for the provision of adequate utilities in Canada will be found in D. W. Smith et al. (1979), while the problems encountered in Northern Europe and North America have been outlined in Smith and Hrudey (1981). A handbook is also available for the design of water supply and sewer systems in the USSR (Federov & Zaborshchikov, 1979; 1982).

WATER SUPPLY

Water is a fundamental requirement for mankind but the quality and quantity required vary according to its proposed use. A safe, adequate supply of potable water is essential for a settlement, while large quantities of water must also be available for fire-fighting or industrial use. This may be water of inferior quality, such as treated wastewater. Thus, the shortage of good quality water can be overcome by re-using some of the water.

In general, finding reliable perennial water sources at low cost is difficult in permafrost regions. Again, the spring or other water source that served a few households when a settlement was first established may be inadequate in quantity or may become polluted as the settlement grows. Good discussions of various aspects of the problem will be found in Carefoot et al. (1981) and Smith and Hrudey (1981).

Sources of Water

Tolstikhin (1940) divided groundwater in permafrost regions into suprapermafrost water, intrapermafrost water and subpermafrost water. The main characteristics of these three categories are summarized in Table 9.1.

Suprapermafrost water is the water which occurs above the permafrost table within the active layer, in taliks or in rivers. It is fed by meltwater, rain, seepages of subsurface water and condensation from humid air. Good examples of springs used as water sources are found in Central Yakutia (Anisimova et al., 1973). Snow and ice can be used as a last resort. Surface sources are particularly susceptible to contamination by man's activities and are usually only available in the unfrozen state for part of the year. They are, therefore, a rather unsatisfactory source of potable water unless rivers or streams can be dammed in a suitable unpolluted catchment. The resulting lake must be large enough and deep enough to contain sufficient water to supply the nearby population throughout the year, in spite of the formation of a thick ice cover in winter. The water supply for Inuvik is an example of a successful man-enlarged lake, although it has to be supplemented by pumping from another lake. Fort McPherson also uses a natural lake but the latter often becomes somewhat polluted by river water during spring floods.

Table 9:1 Classification and Characteristics of Groundwater in Permafrost Areas (modified from Tolstikhin, 1943, and Church, 1974).

Type	Nature of Movement	Temperature	Environment	Yields	Quality
Suprapermafrost					
(a) Seasonal (freezes in winter).	Gravity flow or artesian.	Alternately $>0°C$ and $<0°C$.	Organic material, alluvium, and other unconsolidated material.	Seasonal - nil in winter.	Liable to become contaminated very easily.
(b) Freezes in part during the winter.		Negative.	Alluvium and other unconsolidated material.	Can be large but liable to freeze in pipes if not heated	
(c) Does not freeze in winter.	Gravity flow.	Constantly low negative.			
Intrapermafrost					
(a) Always liquid.	Artesian or gravity flow.	Always positive or always negative.	Alluvium and unconsolidated materials; rarely in solid rock.	Can be good.	Less liable to become contaminated by human wastes. Usually good quality.
(b) Always solid (ground ice).	None.	Always negative.		Very limited and not easily available	
Subpermafrost					
(a) Shallow.	Artesian or stationary; at times sealed by	Low positive or negative if saline	Alluvium, fissures in bedrock, and karst solution channels.	Usually limited.	Liable to have high total dissolved solids. Not readily contaminated by human wastes.
(b) Deep.	Artesian or stationary.	Always positive.	In bedrock, e.g. porous strata, in fissures or in karst solution holes.	Usually finite.	

Deep lakes provide plentiful supplies of water, which may be reasonably warm if collected from below the thermocline, e.g., 2°C at Alert. At Tuktoyaktuk and Coppermine, fresh water is obtained in winter from below sea ice, where it accumulates from the supply from nearby large rivers, e.g., the Mackenzie river.

Intrapermafrost water occurs in taliks within the permafrost, but is not readily detected or mapped. The water is supplied from above or below and is often under an artesian head. Suitable taliks are common below lakes or river beds (Gold & Lachenbruch, 1973), and on the lower slopes of hills in the zone of discontinuous permafrost. The water may be highly mineralized if it represents the unfrozen "mother liquor" of the ice in the nearby permafrost. Pollution by man is less likely than with suprapermafrost water but it is still sufficiently common to cause reluctance by engineers to use this source.

Subpermafrost water is the preferred potable water source, where it is available, e.g., in Alaska and the Yukon Territory. It requires a careful study of water quality and volume prior to development to avoid small water pockets and cryopegs (saline water). However, it is usually safe from pollution by Man's activities and, therefore, is a safer source of water. The relatively high temperature of the water makes it easier to pump compared with water at about 0°C, which has a viscosity about 1.2 to 1.8 higher. Subpermafrost water at Faro, Yukon Territory has a temperature of 2.8-4.4°C. However, there can be problems with high artesian pressures (see Linell, 1973).

Drilling of deep holes can also cause similar problems to oil or gas wells with thawing of permafrost around the hole, followed by collapse or rupture of the pipe (see Chapter 7). At other times, the water may freeze in the pipe if there is insufficient water flow for too long.

Water Storage

When melted snow or ice is used as a last resort in isolated communities in Alaska, it must be stored in insulated shelters in summer and must not become contaminated. Other surface sources such as streams may have periods of low flow and high turbidity. In these cases, water must be collected when of adequate quality and stored until needed. Similarly, a storage facility overcomes the problem of intermittent flow from a deep well and minimizes the risk of freeze-up of the water in the pipes.

The storage facility may consist of storage tanks, dammed-up streams, enlarged or dyked lakes or man-made ponds. Any reservoir must be deep enough to allow for the winter ice cover. Thawing of the adjacent permafrost tends to occur, with consequent slumping of the banks and a tendency to infill the reservoir with silt, e.g., at Chernosevsky, Siberia. The reservoir may also be used to generate electricity. Earth dams have been successfully used at Faro, Yukon Territory and at Chernosevsky by avoiding ice-rich and pervious substrates, by constructing in winter and by using cooling systems such as galleries and cooling pipes (Biyanov, 1965; 1975; Newberry et al., 1978; Biyanov & Makarov, 1980; Melnikov & Olovin, 1983).

Wood, steel or concrete tanks are widely used in Northern Canada and Iceland for small settlements. Above-ground tanks must be insulated or placed inside a heated building. They may be partly buried if the ground lacks ice and special attention must be paid to heat retention to avoid freezing of the water and melting of ground ice by the use of adequate insulation, as well as foundation conditions and icing problems. On ice-rich soils, piles and insulated vented pads will be necessary. The tanks must be vented to avoid collapse due to negative pressure, and freezing of the water must be avoided at all costs.

Water Treatment

It is essential to know the chemical composition of the salts present in a water source as well as its pH, humic acid content and composition. As in other parts of the world, there are specific limits for total dissolved solids (usually around 1000 ppm), as well as for various salts. Epsom salts (sodium sulphate) is the most unacceptable component. The local population can adjust to higher total dissolved solids with time but visitors will encounter problems if the salt content is too high.

pH and humus are particularly critical in permafrost areas and the humus will immediately be apparent due to its brown colour. The organic matter causes corrosion of copper pipes, while iron bacteria utilize the iron and manganese associated with the humus to produce brown flocs of iron hydroxide and manganese oxide. The resulting water cannot be used in laundries nor in the textile and paper industries.

Apart from the aesthetic effects, the humus reacts with halogens to form trihalomethanes, often called THM (Rook, 1974). Thus, when chlorine is added to water containing organic substances, chloroform, bromodichloromethane, dibromochloromethane and bromoform are produced. Peters and Perry (1981) discuss the reactions and show that the yield of THM increases with increasing pH of the water to a pH of 9, increasing time of contact of the halogen, increasing the humic:fulvic acid ratio of the organic matter, increasing molecular weight of the organic matter, and increasing temperature of the water during the reaction.

The primary concern about the THM content is due to the suggested carcinogenicity of these substances. Although humus in water may promote healing of wounds (Biber & Bogolybova, 1952, in Gjessing, 1981), it is also associated with organic and inorganic micropollutants. It is often complexed with substances such as cadmium and can also form complexes with albumen (Visser, 1973). Thus, it can be a vehicle for the transport of undesirable micronutrients and metals into the body.

In 1979, the Environmental Protection Agency in the U.S. announced regulations limiting the permissible level of THM in waste water to 100 μgL^{-1}, based on the mean annual concentration. Other suggested standards are 350 μgL^{-1} in Canada, 25 μgL^{-1} in Germany and 75 μgL^{-1} in the Netherlands. However, there is some uncertainty about the effect of THM on health; the cost of implementation of removal of the organic matter to the prescribed limits would be considerable. It is generally agreed that the risks involved in ceasing chlorination of the water supply are far greater than the risks caused by the THM.

Gjessing (1981) has discussed the main possible treatments to reduce the problems caused by the organic matter, viz.: addition of chemicals to try to reduce the "solubility" of the aquatic humus, addition of chemicals to bleach or mineralize the humus, and filtration with the intention to remove coloured particles and some of the "soluble" colour. Of these, the only generally effective method is the use of filtration through activated charcoal. However, the charcoal must be changed frequently, making this an expensive proposition. Conventional sand filtration only removes 5-10% of the aquatic humus.

Bleaching by ozone results in colour reduction, probably by reduction in size of the organic molecules, but the total organic matter content does not

Table 9.2: Examples of Residential and Community Water Consumption (modified from Armstrong et al., 1981).

Situation.	Location.	Consumption (L/person/d).	Comments.	Reference.
Primarily no permafrost.	Canada.	625	National average for 2780 municipalities, 1975. Mean residential metered (1972-4).	Environmental Protection Service, 1977; Gysi & Lamb, 1977; Howe & Linaweaver, 1967.
	Calgary, Alberta.	217		
	U.S.A.	240	Mean annual residential metered use, Western States, (1960 s).	
Self-haul, permafrost.	Chesterfield Inlet, N.W.T., Whale Cove, N.W.T..	7.5 11.5	Average, no household plumbing	Cameron, 1977.
Trucked, permafrost.	Yellowknife, N.W.T..	118	Mean, trucked, 2.85 persons per house.	Cameron, 1977.
		45-295	Range, trucked, per house.	
Piped, permafrost.	Inuvik, N.W.T.	480-550	Piped circulating.	Smith et al., 1979.
	Ft. McPherson, N.W.T.	250	Piped portion of community.	
	Yellowknife,* N.W.T.	485		
	Whitehorse,* Y.T.	1680	Average annual.	Stanley Associates Engineering Ltd., 1979.
		1135-2500	Range.	
	Dawson City,*#Y.T.	3630-9080	Range.	
	Faro,* Y.T.	790	Average annual.	Cormie, in Armstrong et al., 1981.
	Haines Junction,* Y.T.	570		
	Mayo,* Y.T.	2730		
	Watson Lake,* Y.T.	820		
	Fairbanks,* Alaska	650	Average annual.	Smith et al., 1979.
	Homer,* Alaska	1630		

* Water bleeding in winter for freeze protection.
\# Some leakage.

change much. It causes increased carbon availability to micro-organisms, resulting in increased biological growth in the water unless extra heavy doses of chlorine are added. Unfortunately, this results in maximum production of THM.

Coagulation-flocculation is far more successful. Aluminium sulphate is usually used to produce a white aluminium floc; this has a large surface area which will absorb the negatively charged humus molecules. It is pH-dependent.

Desalinization techniques including distillation, reverse osmosis and freezing have been used in coastal districts of the Arctic. Of these, natural freezing is the cheapest method and is still used to supplement other sources in some communities, e.g., Grise Fjord, Northwest Territories and Barrow, Alaska.

Water Requirements and Transportation Methods

Armstrong et al. (1981) have discussed the water requirements and methods of water conservation in small remote communities in permafrost areas. Obviously the smaller the water use, the more households can be serviced by a given system.

The minimum amount of water for subsistence is about 5.7 L/d for a healthy person (Freedman, 1977). The basic requirement includes drinking and cooking, personal hygiene and laundry.

The average household requires additional water, which can be broken down into toilet (40%), bathing (30%), laundry (15%), and kitchen (13%) and 2% for other uses. These are average figures for cases where water is piped but considerable variation occurs in individual cases.

A community uses additional water for fire-fighting, line flushing, wastage for freeze prevention (bleeding), park watering, street washing and leakage. If there is any industry or commerce present, this will increase the needs still further.

Table 9.2 shows some typical values for residential and community water consumption from North America. The method of transportation of the water is a major controlling factor. Self-haul supplies are essentially subsistence values that can be detrimental to health (A. U. White & Saviour, 1974). Where water supplies are trucked in, conservation techniques are very important due to the high cost per litre involved. The minimum acceptable level of service is 45L/person/d in the Northwest Territories (Cameron et al., 1977), although 70-90 L/person/d is desirable.

Piped systems can and do supply much larger quantities of water. Inuvik and Ft. McPherson are examples of cases where the system is designed to prevent freezing by insulation and heating, although under conditions of limited demand, freezing may still occur. In these cases, the consumption is noticeably lower than where the system is bled to maintain a constant flow. Even in non-permafrost regions, the water use for bleeding can be very high; e.g. at Jasper, Alberta, Associated Engineering Services Ltd. (1975) reported mean annual figures of 270 L/person/d for domestic use, compared with 140 L/person/d for bleeding. The quantity of water used in the bleeding process increases as the winter temperatures become colder.

Water Distribution and Sewage Collection by Pipelines

Improvement of water distribution and sewage collection over self-haul methods involves either trucking or the use of pipelines. Although many small communities such as Teslin, Yukon Territory use trucked water supplies and sewage disposal, the Territorial Governments in Canada are encouraging the use of above-ground (utilidor) or underground water and sewage lines. Heat may be added to the water or to the water mains, and continuous circulation is maintained to prevent freezing. The degree of freeze protection depends on whether the pipes are placed above or below ground. Although the latter is preferred for planning, engineering and aesthetic reasons, the above-ground systems are commonly used in areas of ice-rich permafrost, e.g., Inuvik and Fort McPherson, Northwest Territories, although this is changing.

Underground systems are used in the subarctic climates where at least one month has a mean monthly temperature above 10°C each year, e.g., in Anchorage and Fairbanks, Alaska, Whitehorse, Yukon Territory, Yellowknife, N.W.T. and Gothaab, Greenland. The pipes are protected from freezing by insulation, heating, recirculation of the water and by water wasting (bleeding) from the water line to ensure continuous water flow during periods of low water use. The insulation (5-7.5 cm) is essential to avoid substantial melting of ice-rich permafrost, with attendant ground subsidence. Heating is accomplished by adding water from a warm source or by placing an electrical heating wire along the inside or outside of the pipe's length. This wire must be inside any insulation on the pipe and is 12 W/m in South Greenland and 20 W/m in the north. Lower wattage (8-10 W/m) is used for an electrical wire within the water pipe itself. Where hot springs occur, as in Iceland, the warm water can be carried in large diameter surface pipes for over 20 km to towns such as Reykjavik, Iceland, with only a small loss in temperature.

In Greenland, a unique phenomenon is the summer water line. These are uninsulated pipes laid above ground, which only carry water in the frost-free period (Rosendahl, 1981). These pipes are drained and not used during the rest of the year. This saves money in construction costs but these uninsulated pipes cannot be used in more polar climates.

Above-ground utilidors have been a standard means of providing piped utilities without thawing permafrost in the colder areas (Plates XIX & XX). In many areas, e.g. Greenland, they are being replaced by insulated underground pipes. The utilidors are more expensive because they usually have to be constructed on piles installed in the underlying permafrost and must be adequately insulated and heated to avoid freezing of the pipes. They must be connected to each building and special bridges must be constructed to carry traffic over or under them, e.g. in Inuvik. They cannot readily be disguised and are usually made of wood, aluminium or galvanized steel or concrete. In urban areas of the USSR, e.g. Yakutsk and Mirny, they may be placed below ground to minimize the problem; this method is now being introduced in North America, e.g. at Barrow, Alaska (Zirjacks & Hwang, 1983). In areas with thermally-sensitive, ice-rich permafrost or where excavation equipment is not available, utilidors will continue to be used. They are also normally used for temporary installations. However, where the permafrost is cold ehough, it is possible to use insulated underground pipes, even in icy sediments, e.g. Barrow, Alaska. D. W. Smith and Heinke (1981) and Carefoot et al. (1981) give examples of various types of utilidors. They may also contain central heating, fuel oil, telephone and electricity lines.

Whether above or below ground, piped water systems consist of one or more uninterrupted loops, originating at the pumping station, which must be in a central location. Freeze-up of any section of the loop will halt the flow throughout the loop, so this must be avoided. The water is normally pumped into the loop at 4-7°C and returns at 1-4°C. Where electrical heating of the pipe and good temperature sensors are installed, as in Greenland, the water can be pumped out at 1°C and returned at 0.1°C. Normally empty interconnections between loops and isolating valves can reduce the size of the area affected when a shut-down occurs at a particular location.

Where bleeding is employed to maintain flow, there must be a series of bleeding points in the loop. This can result in water usage as high as 5000 L/person/d. The pipes may be either single pipes to large water users, e.g., apartments, or dual pipes with a large supply line and smaller (5-15 cm) return line. However, the single pipe loop is preferred for most applications. Heat tracing lines are expensive to install and operate but are the main method of avoiding freezing of the pipes. Steam and hot water thawing lines (usually using hot water from power generators) may be used to thaw frozen pipes.

The sewer system is usually of the conventional gravity type, although vacuum or pressure systems can be used if absolutely necessary. The pipes are insulated and storm water is excluded from the system to conserve the heat within the waste-water leaving the buildings and to avoid melting the ice in the adjacent permafrost. Water bled from the water lines alters the sewage quantity and quality, as will be described below. D. W. Smith and Heinke (1981, p. 10) give the following comparative construction costs excluding costs of excavation: buried utilidors, $400-$1200/m; on-surface utilidors, $200-$300/m; above-surface utilidors, $600-$1700/m; buried pipelines, $100-$300/m; and on-surface pipelines $50-$100/m. This is the reason for the slow improvement in water distribution and sewage collection systems in the smaller northern communities in permafrost regions.

WASTE DISPOSAL

Wastes can be divided into two broad groups, viz.: wastewater and solid wastes. Both can cause problems of disease transmission or damage to the environment if not dealt with properly. Currently used methods of treatment and disposal are often relatively primitive and are being upgraded (Heinke, 1974; Slupsky, 1976; Rosendahl, 1981; Balmer, 1981; D. W. Smith & Heinke, 1981). Good bibliographies of methods will be found in Snodgrass (1971) and Cameron and D. W. Smith (1977).

Wastewater Treatment and Disposal

The nature of the water distribution system largely decides the wastewater characteristics (Smith & Heinke, 1981). Community wastewaters can conveniently be divided into four categories on the basis of the concentration of contaminants, i.e., undiluted wastes, moderately diluted wastewater ("black water"), conventional strength wastewater (also called "grey water") and very dilute wastewater. Usually a community produces wastewater of at least two concentrations and each category requires a different treatment.

Undiluted Waste

This is the product of the use of bucket toilets, which are commonly used in small settlements where the cost of a piped system is prohibitive. Ideally, the "honey-bags" are collected and trucked to a special disposal pit as far away as possible from the community and located in an area which is not used for other purposes. The hydrology of the area should be checked to ensure minimal risk of contamination of the water supply (Heinke & Prasad, 1977). The size of the pit should be at least 0.5 m^3 per person per year. It should be separate from the landfill and covered in and treated with lime when full. It acts as a giant holding tank; laboratory experiments show that no decomposition is occurring and that the pathogens may be viable for many years (Prasad & Heinke, 1981). Anaerobic digestion is possible but uneconomic. Along coasts, e.g. Greenland, the wastes are dumped into the sea (Rosendahl, 1981).

Direct incineration is not used at present but incineration of sludge from other treatment processes (Eggner & Tomlinson, 1978) and of moderately concentrated wastes has been practised. The main problems seem to be the expense and the difficulties connected with the operation of such a facility on a small scale in a remote settlement.

Moderately Diluted Wastes

These are produced by using holding tanks, pressure sewers and vacuum sewers. Conventional anaerobic digestion is feasible and practical at a temperature of 20°C with a 30-day minimum detention time (Prasad & Heinke, 1981). Digestion at lower temperatures requires that a higher percentage of methane bacteria be used in the reactor, together with a longer detention time. The most promising method appears to be an anaerobic filter treatment process, although both the effluent and primary sludge require further treatment.

In the past, methods of treatment have included lagoons (both aerated and faculative), oxidation ditches, extended aeration plants, attached growth devices and physico-chemical processes. These work best with more dilute wastes. The faculative lagoon is particularly common due to its low capital and operating costs and simple maintenance requirements (Grainge & Dawson, 1969). Allowance must be made for ice cover so that the lagoon, together with its inlet and outlet, will not freeze. Detention times need to be 8-12 months.

Aerated lagoons require much better maintenance, which is usually not available but oxidation ditches have been very successful (Murphy & Rangarathan, 1974). Attached growth systems need protection from freezing and are, therefore, not much used.

Physico-chemical methods have been used at work camps, especially along the Trans-Alaska Pipeline and at dam construction sites near James Bay. They are very costly, labour-intensive and require special chemicals and encapsulation to prevent freezing. However, they take up less space, have consistent effluent quality and minimum impact on the site. They are too expensive for normal long-term use by settlements.

Conventional Strength Wastewater

Hrudey and Raniga (1981) discuss the characteristics of conventional strength wastewaters. Such waters are produced by piped systems and are characterized by wide fluctuations in flow, organic loading and content of various toxic chemicals. The content of microorganisms is high, indicating that disinfection must be part of the treatment process. Shigellosis (bacillary dysentery) and hepatitis type A were 73.9 and 21.1 times as common in the Northwest Territories as in the rest of Canada between 1975 and 1979, inclusive, indicating the risks involved without disinfection.

Apart from disinfection of the effluent, conventional strength wastewaters can be treated as in the case of moderately diluted wastes. They are rich in nutrients and the effluent can, therefore, provide much needed nutrients to the soils. Balmér (1981) discusses the Swedish experience with treatment of these wastes.

Very Dilute Wastewaters

When bleeding of the water line is used, the excess water is bled into the sewer lines and produces very dilute, low temperature wastewater. This represents a waste disposal problem involving relatively large volumes of effluent which must be kept from freezing in the sewage line. Often heat must be added to the pipes at one or more locations, e.g., at Kotzbue, Alaska. Type of pipe (above or below ground), depth of burial, amount of insulation, use of protective outer layers, heat loss during treatment and disposal and the use of back-up freeze protection devices all affect the total heat loss. User heat addition or subtraction and the temperature of the raw water are also involved. At Whitehorse, a controlled blend of 1-2°C surface water and 4°C groundwater is sufficient to avoid freezing.

D. W. Smith and Given (1981) discuss the various types of treatments used in Canada. The method is usually selected according to the requirements of the regulatory agencies for purity of the effluent. Where low BODs and suspended solids are required, a short-retention lagoon and screening are employed, e.g. at Hay River, Ft. McPherson and Inuvik, Northwest Territories, and Whitehorse, Yukon Territory. The performance of the lagoon at Inuvik is described by Miyamoto and Heinke (1979). BOD is reduced by 30% in winter and 90% in summer, while 80% of the suspended solids are deposited as sludge. The effluent is discharged into the east channel of the Mackenzie river.

Zero discharge of pollutants in winter is needed in some situations and involves the use of a long retention lagoon, as at Watson Lake, or the Imperial Oil lagoon at Tuktoyaktuk, Northwest Territories. It requires a much larger capacity lagoon. In both cases, the effluent will still contain pollutants. Water depths must exceed 2 m to avoid freezing to the bed in winter. Smell from the sludge can be a problem.

Where more stringent controls on effluent quality are applied, e.g., in the United States, a secondary treatment plant such as a rotating biological contactor is necessary. This is very expensive because of the very large water volumes involved. Accordingly, in Alaska, the system of bleeding the water lines tends to be reduced or eliminated so as to make the secondary treatment more economic. In northern Canada, mechanical treatment plants are only found at a few locations, e.g., Churchill and Thompson, Manitoba, and Carmacks, Yukon Territory.

Solid Waste Disposal

Until recently, this was the most neglected area of sanitation in the Arctic. Waste was usually stored in the house in 180 L oil drums. Average waste production may be 1-1.6 kg/person/d, depending on whether the garbage is burned prior to disposal. Refuse is usually collected once or twice a week and preferably kept separate from the liquid wastes. Unless the trucks are covered, some garbage blows away.

The garbage is usually placed in an open dump. For much of the year, it cannot be covered easily with soil, while low temperatures minimize degradation of the material, which is effectively in cold storage (Straughn, 1972). Modified landfill on slopes or in trenches is preferable, where possible. Wind tends to blow the uncovered material across the landscape and there can be water pollution if the site is not chosen with care.

Large communities such as Fairbanks, Alaska, can afford to use shredders, baling and incineration, but these processes are expensive. Without an automobile crusher (used at Fairbanks), a place must be found for discarded machinery, automobiles, snowmobiles and appliances. Where these can be crushed, they can be placed with the rest of the solid waste.

ELECTRICITY

An essential part of survival in the north for communities and industry is the provision of suitable power and communications systems. Until recently, these shared the need for transmission lines, although now telephone lines are becoming redundant. Even so, repeater towers are a common feature along the main transportation routes. The supply of power must be as continuous as possible, with a minimum of maintenance. Distances are great and equipment for repairs may not be readily available.

There are three aspects of the provision of electricity that merit special attention in permafrost areas, viz.: the building and operation of dams, the foundations of transmission towers, and the provision of adequate grounding of electrical equipment.

Hydroelectric Dams

Power for northern communities can come from one of three sources, viz.: natural gas, oil, or hydroelectricity. Oil and gas are available in reliable supplies only at a few sites such as Norman Wells, Northwest Territories, and Yakutsk, Siberia. Supplies of gas and oil can be moved by boat in summer but this still leaves large tracts without suitable supplies. For these, hydroelectricity has to be used.

This must usually be generated by damming up a river whose flow regime is suitable. It must not overflow the dam and the temperature of the water needs to be close to 0°C to enable the dam to remain frozen. The dams are almost invariably earth-filled and a considerable number have been successfully constructed in Siberia. Careful investigation of the local conditions is essential to avoid problems (Bogoslovskiy et al., 1966; Sayles, 1984). Often the construction must be staged to ensure that the material in the dam is suitably frozen and icy.

In areas of continuous permafrost, ground temperatures will be low enough to allow short impervious frozen dams to be constructed by placement of fill in thin layers in winter (Fulwider, 1973). However, for larger dams, artificial freezing (mechanical refrigeration systems, circulating cold air) is used to construct and maintain an impervious frozen core and foundation (Biyanov, 1975). All unfrozen material below the river, overlying competent bedrock or compact frozen till, must be removed. At Vilyui hydroelectric dam at Chernosevski, flood waters were allowed to pass over the partly constructed dam to add ice to cement the material. However the structure is constructed, it is essential to carefully monitor the thermal regime, the settlement and performance of the structure (G. H. Johnston, 1965; Tystovich, 1975; Gupta et al., 1973) to provide adequate warning of any necessary remedial work. Settlement as icy permafrost thaws may be considerable.

The main problems in the zone of continuous permafrost include expansion of the talik under the man-made lake, the slumping of thawed masses of earth into the lake, thawing around the overflow channel, and thawing of the underlying bedrock and the dam. G. H. Johnston and MacPherson (1981) give a more detailed summary of these problems, while Biyanov (1975) and Biyanov and Makarov (1980) summarize the experience in the USSR.

In the zone of discontinuous permafrost, dams must be designed for thawed conditions. Impervious dams are normally used to reduce seepage losses, although the exact design depends on the nature of the bedrock, the type of fill available and relative cost of the various design alternatives (MacDonald et al., 1960). Care must be taken to avoid deterioration of the core (Duguid et al., 1983) and new dam material may have to be added to compensate for the inevitable settlement as icy material thaws. Ice-rich, fine-grained material is unsatisfactory unless its ice content is reduced. Thaw settlement progresses downwards under the lake and inwards into the impermeable dam in a stepwise fashion under the influence of the seasonal temperature fluctuations superimposed on the continuous effect of the water body (G. H. Johnston, 1969). Grouting may be necessary to fill voids developing in the underlying bedrock by melting of ice. This is normally done from galleries within the dam. Melnikov and Olovin (1983) describe the permafrost dynamics of the Vilyui river hydroelectric scheme.

Tower Foundations

Transmission lines are needed to move the electricity to various communities. They must be built across very varied terrain underlain by permafrost. The tower structures will be subjected to high winds, the weight of the wires, and ice loads. In some areas, there may be risk of earthquakes. If icy permafrost starts to thaw, differential settlement may occur. This is most likely in the warmer regions near the margins of the permafrost zone. Buried lines are avoided because frost heaving and thaw settlement tend to cause breaks in the cables.

A variety of techniques that are available are summarized in Davison et al. (1981, pp. 331-341). The simplest system consists of stub piles or wooden posts augered into the permafrost. The poles are then strapped to the wooden piles. This works reasonably well for power distribution posts and for telephone wires. Where problems occur with frost heaving or thaw settlement, the poles may be set in rock-filled cribs.

Major transmission lines and microwave towers require more substantial foundations that are stable. Four-legged, free-standing towers or steel-guyed supports are normally used since they will resist uplift, horizontal and downward stresses. For transmission lines, timber, reinforced concrete and steel footings are usually adequate. Piles are usually used for communications towers (e.g. Nees, 1951) but are more expensive. Footings and grillages are placed in the permafrost well below the anticipated new depth of the active layer. Disturbance and conduction of heat downwards by the steel will cause the active layer to thicken after construction. Insulation is commonly placed around the steel and the footings to attempt to minimize the changes in thermal regime.

Anchorages for guy wires must withstand stresses from sudden wind gusts (Crowley, 1977). Design is chosen to allow for inevitable progressive creep in icy permafrost, which must be kept within the allowable tolerance for the life of the structure. The three main types are block anchors, piles and grouted rod anchors. Block anchors are heavy weights or blocks, preferably buried in the permafrost, which are relatively free of creep. If they are placed in the active layer, they will be subject to frost heave and thaw settlement. Piles are less able to withstand vertical loads and, like rod and screw anchors, must be placed deep in the permafrost to be effective. Load tests in the field are advisable to establish the probable performance of the anchors (see Johnston & Ladanyi, 1972, 1974).

Electrical Grounding

Frozen ground has a very high resistivity (Hessler & Franzke, 1958) and it is, therefore, difficult to obtain a good ground connection in areas of continuous permafrost. This creates problems in combatting lightning strikes and equipment failure. Inadequate grounding can cause loss of life and injury and requires special precautions.

The best ground is obtained by attaching the negative cable to steel water pipes that are buried underground. Lakes or wet, unfrozen ground will permit normal grounding as long as the ground wire is led into the associated talik. Cased water wells or streams also present good potential grounding locations. The quality of the ground will tend to vary with the season, being worst in the spring when the unfrozen zone is smallest. Least problems are encountered in areas of sporadic permafrost and the problem increases towards the zone of continuous permafrost. As a result, transmission lines in areas of continuous permafrost tend to be located in lowlands near lakes and rivers that do not freeze to the bottom in winter. Particularly bad grounding will occur where the ground becomes very cold in winter and permafrost will be very thick, e.g., on road surfaces.

Multigrounding is commonly employed on transformers and other key installations, and tests for adequate grounding of electrical equipment and towers must be made under both summer and winter conditions.

Chapter Ten

AGRICULTURE AND FORESTRY

The indigenous peoples of the permafrost areas practised relatively limited agriculture. They were primarily hunters, fisherman and herders, as is seen in Tibet today. When the Russians expanded their territory northwards and eastwards in the sixteenth century, the first products from the new lands were furs. Russian settlements in Alaska were concentrated along the south coast and there was relatively little travel inland except by trappers. In Asia, the indigenous peoples were involved in hunting and herding; horses were widely used for transport. In North America, the Inuit relied primarily on fishing and hunting. It is only in this century that reindeer herding was added to their skills.

With the increased European expansion into permafrost regions, the newcomers attempted to produce the vegetables and field crops that they were accustomed to before they moved. They needed timber for building and for keeping warm during the long winters. As a result, new crops were introduced and the indigenous people started to learn from the limited successes of the newcomers. It is from these beginnings that the present pattern of agriculture and forestry in permafrost areas has developed.

PERMAFROST, AGRICULTURE AND FORESTRY

Permafrost is the result of climatic conditions which are restrictive to most trees of commercial value as well as to all but the most hardy agricultural and horticultural crops. Permafrost does not necessarily cause the main restrictions but it does modify soil conditions very considerably. In addition, alteration of the vegetation cover inevitably alters the thermal regime of the ground, and usually causes degradation of part of the permafrost and the development of thermokarst phenomena in ice-rich sediments (Péwé, 1954; 1966a, 1966b; Solov'yev, 1962, 1973a, 1973b; Czudek & Demek, 1970). The resulting soil disturbance is disruptive to the vegetation.

The limitations imposed by permafrost include a shallow active layer, a perched water table on the frozen substrate, slow thawing of the icy ground and the resultant cool ground temperatures. In free-draining substrates such as sands and gravels, the perched water table will be advantageous during dry, hot spells since it will provide an adequate water supply for the vegetation, e.g., on the sandy deposits near Yakutsk. Unfortunately, many sediments are fine-grained and consequently need artificial drainage to improve aeration and to increase the rate of warming of the soil. This can only readily be provided near river courses. As a result, the potential agricultural areas of northern North America and the USSR tend to be concentrated along the major rivers (Figs. 10.1 & 10.2). However, the low-lying swampy muskeg of much of the Northwest Territories and northern Quebec presents too great a problem to drain and use.

Where the permafrost is not particularly ice-rich, removal of the vegetation permits deepening of the active layer so that the soils warm more quickly and become better drained. On coarse-grained soils at well-drained sites in areas of low summer precipitation, irrigation may even be necessary, e.g., in the Matanuska valley (Alaska Geographic, 1984, p. 69). Sprinkler systems are normally employed to avoid problems with erosion and because the water droplets warm considerably during their flight through the air.

The most effective method of irrigation may be trickle irrigation since the water is applied near the plant and so less water needs to be applied.

The effects of melting of ground ice to produce thermokarst were described in Chapter 3 but they are devastating to agriculture and to horticulture. Once collapse structures begin to form, it is hazardous to venture onto the ground (Péwé, 1954; 1966a & b) and tractors and other machinery can fall into the pits. Microrelief with an amplitude greater than 1.5 m is very difficult to cultivate. It is only when the ground ice has melted out completely and the resultant water has drained away that cultivation can safely be undertaken. However, the end product, the alas depression, is one of the main reasons for the unusual success of agriculture around Yakutsk in Siberia.

CLIMATE, AGRICULTURE AND FORESTRY

The same harsh temperature regimes that produce the negative ground temperatures typical of permafrost create severe limitations for all but the most hardy plants. Thus in central Siberia, the Siberian pine only extends as far north along the Lena river as its junction with the Vilyuy, while the northern limits of the Siberian spruce lie in the Verkhoyansk mountains. Thus only the Siberian larch, birch and

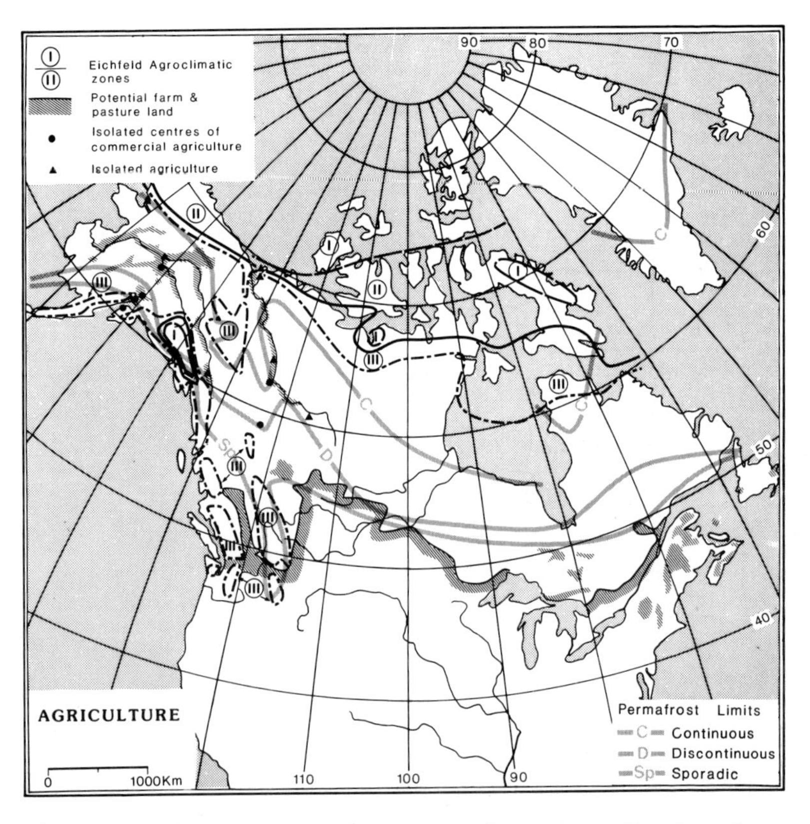

Figure 10.2: Potential Farm and Pasture Land and Isolated Centres of Agriculture in North America in Relation to Permafrost and the Eichfeld Agroclimatic Zones.

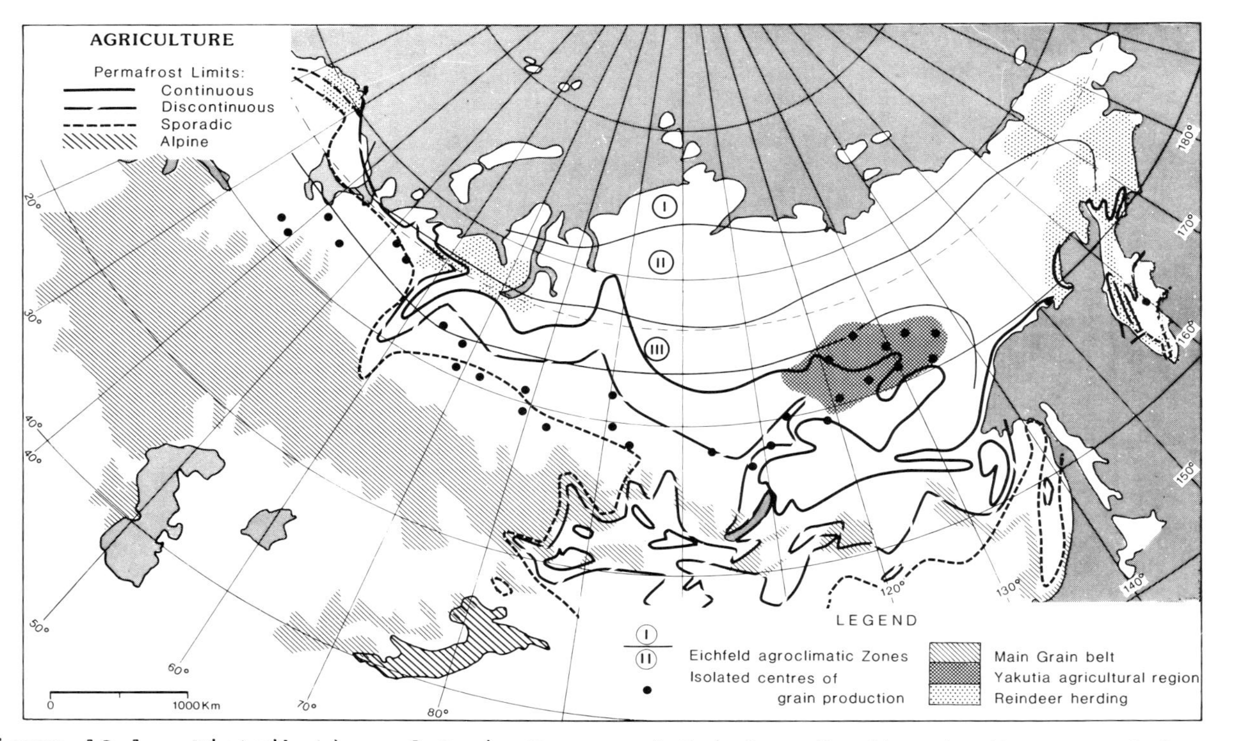

Figure 10.1: Distribution of Grain Crops and Reindeer Herding in the USSR (after USSR, 1980, p. 163; Allen,1985; Gregory,1968), in relation to the Agroclimatic zones of Eichfeld (1931).

balsam poplar extend northwards to treeline (Gregory, 1968).

In spite of this, growth of crops and rearing of animals is still possible and the results are used to improve the diet of the few settlements in the region. Logging can be carried out, although the quality of the timber is poor near treeline and the rate of regeneration is slow.

To study the problems of agriculture in the north, agricultural research stations were established in several locations in Siberia, Canada and Alaska. The first experiments in Siberia were the pioneer research of Lomonosov in the eighteenth century, and were continued under the auspices of the Geographical Society of St. Petersburg in the nineteenth century, often by Polish exiles such as Czerski, Czekanowski, and Dybowski (see Stanek, 1970). The first true experimental station in Russia was that at Khibiny in the Murmansk district, while a string of seven agricultural experimental stations was established throughout southern and central Alaska following the visit of Walter H. Evans in 1897 (Gasser, 1948; Alaska Geographic, 1984, p. 18). In Canada, agricultural experimental stations were established at selected northern settlements such as Norman Wells and Fort Simpson, but elsewhere, much of the research was carried out by trial and error under the guidance of the local priest.

Eichfeld (1931) divided northern Eurasia into three agroclimatic zones (Fig. 10.1). In the coldest zone (I), vegetable cultivation is limited to hot beds and hot houses. Hydroponics are often used. It represents the tundra region where the mean temperature of the vegetative period is 2.6-5°C. In zone II, field culture of early vegetables such as potatoes and cabbage is possible. Mean temperatures during the vegetative period average 7-8°C and both zones II and III represent the northern relatively open-canopy forest of stunted larch, birch and balsam poplar forest. Zone III permits the growth of vegetables as a field crop together with early ripening cereals. The mean temperature during the vegetative period is 9-10°C and the southern boundary approximates the northern limit of the denser spruce-larch forest with good usable timber.

Southwards of these zones is the main Taiga forest zone in which the trees are densely spaced and grow very tall, given adequate soil moisture. It represents the main forest reserve of the Soviet Union. Factors such as precipitation, distribution of ground ice, nearness to markets, etc. become the dominant limitations to agriculture.

The North American situation is a little different. Eichfeld's system has not previously been applied here, probably because population pressure is much lower and other limiting factors are involved. There is far less land suitable for agriculture (Fig. 10.2), partly due to mountains of the northern Cordillera in the west and partly due to the rocky surface of the glaciated lowlands of eastern Canada and the Canadian Shield which have extensive peaty swamps in the low places, marked by stunted black spruce forest. An agroclimatic subdivision using the three zones of Eichfeld and based on data from Environment Canada (1982) is also plotted in Figure 10.2. It will be seen immediately that the zones form a belt encompassing the Arctic Islands. The zones also occur in the Yukon, the coastal ranges of Alaska and British Columbia, and in western Newfoundland. However, extreme cold is less of a factor in limiting agriculture in mainland North America than in Russia.

GENERAL METHODS OF EXTENDING THE RANGES OF CROPS POLEWARDS

A good, recent discussion will be found in Alaska Geographic (1984), while additional references include Brown (1970a); Gasser (1948); Hunt (1978); McCracken and Revel (1982); Nowsad (1963); and Department of Agriculture (1970, 1977). Bliss (1978) has discussed the agricultural potential of the north, while adaptations of plants to cold regions are discussed in Bliss (1962). The effects of man in altering his environment in the north are discussed by McTaggart Cowan (1969) and Klebesadel and Restad (1981).

Vegetable production can be carried out intensively, using relatively little space in contrast to livestock production. The latter must be provided with adequate fodder for the long winters and, hence, is more difficult to accomplish.

Vegetable Production

Successful vegetable production outdoors requires careful selection of the site. This should take into account insolation, snow cover, wind angle and force and distance from frost pockets resulting from cold air drainage. South-facing slopes receive maximum insolation and near water-bodies, can take advantage of light reflected from the water surface. Wind-breaks are commonly left on all sides except downhill, so that wind is minimised but cold air can drain away. Where trees are not available, fences are used.

The presence or absence of permafrost is very important. In the discontinuous zone, soils without permafrost are chosen since they are warmer. In areas of continuous permafrost, it is necessary to clear the land so as to thaw the upper zone of soil. The gradual recession of the permafrost table may take five years, and soils with ice wedges and large ice bodies should be avoided. Thawing deepens the active layer but the melting ice in the permafrost usually creates drainage problems at first. Then, after the active layer has thickened, the drainage will be satisfactory. Not all the permafrost disappears; it merely reaches a new thermal equilibrium, providing a much deeper active layer.

Sometimes where soils are shallow, soil is brought in and compost is added to it so as to form upstanding beds, about 50 cm high, supported by rocks or wood, e.g. railroad ties; loamy soils are preferred since clay soils require artificial drainage, while sandy soils require supplemental irrigation. Vegetable wastes are collected, composted and then added to the soil to improve its quality. Frequent cultivation keeps down weeds. Acid soils will need liming and yields are usually improved considerably by the use of fertilizers. However, distance from sources of fertilizers and lime can make this too expensive.

Artificial warming of the soil helps in giving vegetables an early start. Good drainage makes a 2°C difference in the rate of warming, while the use of a clear polyethylene plastic cover can warm the ground another 5-20°C, depending on local conditions. The conventional straw, black plastic, sawdust and newspaper mulches of lower latitudes prevent the ground warming quickly and, hence, must be avoided. Ridging of the soil to form low, flat-topped mounds at least 50 cm wide is a common technique for maximizing warming by the sun. Clear plastic mulches are even more effective than polyethylene sheeting at prewarming the soil. At Circle Hot Springs, Alaska, the hot water is used to heat the ground by allowing it to flow through buried underground pipes. Fertilising and warming by irrigation of the ground with warm waste waters from industrial plants has been tried in other areas (Dodolina, 1978; Kovaleva et al., 1978; Palazzo, 1976).

The crops need to be hardy and potatoes, carrots, red beets and cabbages are the most common vegetables in Russia. In fact, a very wide range of quick growing vegetables as well as many berry crops will yield well in the north (see the listing in Alaska Geographic, 1984). Where possible, annuals are started indoors or in cold frames so that they can be gradually hardened and then transplanted into the garden as soon as the soil is warm enough.

Greenhouses are widely used to provide a reliable climate, particularly for commercial and large scale production. The common type is the summer greenhouse, used to grow crops during the period of long summer days. At these latitudes, total insolation reaching a horizontal surface during the growing season differs little with latitude, other things being equal, because the lower angle of incidence is compensated for by the length of the day. Free-standing glasshouses are better than lean-to types since they make use of the total insolation. They need careful ventilation and management but effectively increase the length of the growing season so that a wider range of vegetables and varieties can be grown. In general, these greenhouses are not heated, except where they are close to hot springs or power plants.

The winter greenhouses of the Soviet Union are greenhouses used for year-round production of crops. They are extremely expensive to operate unless they are near power stations and are designed to function at air temperatures as low as -50°C (Wein, 1984). Artificial lighting is necessary to augment the insolation but the greenhouses can provide large quantities of perishable vegetables for isolated settlements that could not be obtained from elsewhere except by air, which would be even more expensive.

Water control is critical. Too much water slows the warming process and reduces yields, whereas too little water results in poor growth. Where irrigation is used, a warm water source is desirable and it should be in the form of gentle sprinkler or trickle irrigation to avoid soil erosion. Enough water must be applied to soak the soil so that the plant roots develop through a deeper soil zone. Sprinkler irrigation is used where water is available under pressure.

Drainage can be achieved by terracing, by making raised beds, adding sand, compost or peat, or by using drainage ditches or pipes. However, pipes tend to get broken and heaved out of the ground by frost in silty soils, so that ditches are more commonly used, e.g. in southern Iceland.

Livestock

Livestock production is of two types, viz.: the extensive farming operations requiring large areas of land to produce forage crops, grain, and hay, and the cottage livestock industries involving poultry, rabbits and hogs. A good summary of the relative success of various types of livestock in Alaska is given in Alaska Geographic (1984). Sheep are the best adapted to the conditions but need winter feed, fresh water daily and shelter from the wind. They are the main form of livestock on farms in Iceland.

Dairy cattle are very important since they are the only available source of fresh milk. The dairy cows need a controlled environment although the calves can be raised outside in a sheltered enclosure. They require fresh water daily and large quantities of hay, silage and high-protein concentrate. Thus, they can only be successfully raised where there are extensive hay fields and where some grain can be grown near large settlements. Thus, in North America, the Matanuska and Tanana valleys have dairy farms supplying Anchorage and Fairbanks respectively. Elsewhere, beef cattle consisting of a variety of breeds can be successfully produced if there is enough feed available. They need protection from wind but not a controlled environment.

Hogs are best produced in total confinement. They require a mixture of feed barley and vegetable protein products for best results, so successful commercial operations are confined to the grain-growing centres.

Goats must have winter feed including hay and silage and also shelter from the wind. They are found in small groups and as individuals, reared in settlements throughout the permafrost region. Other animals typical of the cottage livestock industry include poultry (chicken, ducks, geese and turkeys), which must have a proper enclosure and an even temperature, and rabbits. The latter help dispose of vegetable wastes.

Herding and Harvesting of Indigenous Animals

In Eurasia, reindeer herding is a well-established form of agriculture, which takes place in large areas at the limit of northern agriculture. The indigenous peoples learnt to conserve and harvest this resource long ago.

The natural carrying capacity of hunters of caribou in North America has been estimated at one person per 650 km^2 (Vallee, 1967). Once the population of the north exceeded this, a decline in numbers of caribou began (McTaggart Cowan, 1969). This has recently been accelerated by greater efficiency of transportation of men using new roads, snowmobiles and planes. Musk-ox have shown a similar decline.

In Russia, the herds of the indigenous species of wild animals have also been greatly depleted, so the reindeer herds are the main significant source of meat from indigenous species.

A natural development was to introduce reindeer into North America to try to alleviate the problem of over-hunting of the wild caribou. McTaggart Cowan (1969), Krebs (1961), and Roberts (1942) discuss the results. In general, the experiments may be termed failures since the local people did not adapt to or quickly learn the potential and the ecological requirements of the domesticated reindeer. Although Porsild (1929) estimated the carrying capacity for reindeer between Coppermine and Alaska at about 500,000 animals, only a limited number remain. These are operated by the Inuit at Tuktoyaktuk. Both the musk-ox and the reindeer have the potential for a ranching industry.

AGRICULTURE

This is best discussed by region, since each has had a different history and has different environmental problems.

USSR

The main grain-producing regions of the USSR lie south of the permafrost zone, with the exception of rye. The latter is found growing as an extensive and important crop in the lowland areas of discontinuous permafrost (Fig. 10.1). It is associated with significant but limited pockets of cultivation of barley, oats and buckwheat. Summers are very hot and there are over 90 frost-free days, so there is adequate time for the production of good crops of the more hardy strains of these grains. Locally, north of Vladivostok some rice is produced.

Within the main permafrost zone, the land is mainly under forest. However, particularly along the Lena, Yenisey and Ob rivers, there are a series of isolated local centres of grain production where a combination of good soil conditions and hardy strains of fast maturing crops have favoured the development of local settlements producing significant quantities of grain for local consumption. Areas of alas valleys along the Aldan river near Yakutsk are examples where the ground ice has melted and widespread cultivation is possible without causing further problems with thermokarst. Gregory (1968, pp. 595-601) provides a concise history of the evolution of agriculture in this area.

Further north, within the agroclimatic zones II and III, isolated pockets of cultivation are found near settlements along the main rivers. In zone III, grain production is commonly practised in European Russia, but vegetables, grazing land and hayfields are dominant in zone II. Major areas of reindeer herding straddle the boundary between Zone II and Zone III, particularly in Northeast Russia (Gregory, 1968).

In these northern areas, 85% of the cultivated land is used to provide sustenance for sheep and cattle. The object of agriculture is to provide local supplies of perishable foods such as milk, potatoes, vegetables and meat (Wein, 1984). These are augmented by food brought in from designated areas in the south along the main transportation routes, viz.: roads, rivers and railways. Wein (1984) provides a good review of the present situation based on a number of Soviet sources, including Boyev and Gabov (1981). It is hoped to increase local production considerably in the next five years, with particular emphasis on milk production (Shabad, 1982). This will entail extensive reclamation work on potential grazing land by removing the existing vegetation, draining the land and reducing its acidity.

Between 1965 and 1977, greenhouse complex production near the larger popularion centres increased vegetable production by 25% in West Siberia, 29% in East Siberia and 32% in the Far East. Of the greenhouse complexes, 61% were summer greenhouses in 1978, the rest being operated in winter. These are being increased in area in approximately the same proportions during the next five years (290 ha. to 110 ha.), but they are trying to locate the new complexes by power plants. This reduces energy costs by at least six-fold.

Special-purpose state farms are being developed in the south to try to increase production. Further north, the sparse population is being encouraged to grow vegetables both at their urban homes and around their summer cottages (dachas). The produce is used by the family, with the surplus being sold at farmers' markets.

The opening of the BAM railway in 1984 has resulted in the movement of people to this new frontier (Wein, 1984; Druzenko et al., 1984). This is an area of continuous permafrost that is being developed and settled. The forest is being cleared and agriculture is being established. The length of the growing season is adequate for production of grain, but it remains to be seen whether problems with thermokarst can be avoided.

North America

North America differs from Siberia in having very restricted areas with suitable soils for agricultural pursuits (Fig. 10.2). The largest areas are in Alaska and Yukon Territory, with a second area along the Mackenzie valley.

When Alaska was settled by the Russians, they started to grow their own food so that they would not have to eat fish all the time. Unfortunately, grains would not usually ripen in the wet coastal climate, so they were generally without flour and bread. Fortunately, vegetables were more successful, notably tubers such as potatoes and rutabagas, while lettuce and cabbage grew well but were watery. Turnips, radishes, beets, carrots, onions and garlic did well.

The promyshleaniki also brought with them a few farm animals such as chickens, cows, sheep, pigs, horses and goats. The meat of the chickens and pigs generally tasted like fish, due to the fish fed to the animals. European travellers from the east coast tended not to bring seeds and animals when they travelled into the north.

The main expansion in agriculture in Alaska occurred as a result of the gold rushes. These opened up new areas and it was far cheaper to grow food locally than to import it. The United States set up seven agricultural research stations at the turn of the century to carry out experiments to find the most successful methods of agriculture in the various regions of central and southern Alaska. They developed good varieties of many crops, but it soon became apparent that the only two places where commercial agriculture was both potentially successful and required were in the Matanuska valley near Anchorage and in the Tanana valley near Fairbanks. The other stations were, therefore, closed and the main research station moved to the Matanuska valley.

Vegetable gardens and some livestock can be seen at most small settlements throughout Alaska and much of the southern Yukon Territory. However, it is now cheaper to import vegetables from California or British Columbia by sea, road, or air, than to grow them locally. As a result, agriculture is not nearly as well developed as it could be.

In Canada, the early agricultural efforts tended to be at the missions, where the local priest encouraged the development of gardening. Some experiments were carried out at Norman Wells and Fort Simpson before World War II. The main development of the Mackenzie corridor to Tuktoyaktuk occurred during and after the 1950s, and garden-type agriculture is carried out in the new settlements. However, the barge-traffic along the Mackenzie river followed by the construction of the Dempster Highway made the importation of produce from the south relatively easy.

Further east, there are virtually no areas of northern settlement with the combination of adequate climate and suitable soils to permit northern agriculture.

The Tibetan Plateau

The areas of the Tibetan Plateau which are underlain by permafrost lie in an arid region. Meltwaters from the snows and glaciers of the adjacent high mountains provide the main sources of fresh water, but these often end in saline playa lake beds (Kowalkowski, 1978). The adjacent soils are saline and alkaline Brown Desert soils, with active pingos.

The vegetation consists of a 30% cover of purple-flowered needlegrass. Zhang et al. (1982) report yields of wet, fresh-cut grass of only 300 kg/hectare. It, therefore, requires considerable effort to obtain a reasonable supply of hay for the winter. However, the strong solar radiation speeds up photosynthesis, so that the grasses are of high nutritional value. They may contain up to 16.8% coarse protein. The grasses grow and mature earlier at lower altitudes, so a continuous supply of grasses suitable for hay-making can be found throughout the summer, provided that the Tibetans pursue a form of transhumance up the mountain sides.

The main form of agriculture is extensive grazing using the indigenous animal, the yak, and sheep (see National Geographic Society, 1982b). The yak is an important beast of burden, working alongside horses, mules and goats. The yak also provides a rich milk (6-8% butter-fat) used for making butter and cheese, while yak hair is used to weave the material for the nomads' black tents; yak dung is employed in starting the fires used to keep the families warm during the cold nights. Yak meat is the major source of protein and is often dried and smoked for use in winter. Yak skins are used to make boats and boots. They can be traded for tea and grain grown in the lower elevations in the Himalayas. The yaks can also be used to draw ploughs.

The sheep are grazed on the high summer pastures and are brought down with the yaks to pens close to the family tents in winter. There the animals subsist on hand-cut hay, carefully collected and dried during the summer. The sheep yield mutton, wool and skins. The wool is woven into cloth during the long winters to provide the necessary warm clothing. In summer, shepherds weave the wool together with plant fibres into a material used for the soles of their boots. However, the wool cannot be used for spinning.

The yak is highly resistant to the cold climate, grazing in the open at temperatures down to -30°C. However, they cease to eat and start gasping for breath at 15-20°C. Both sheep and yaks are adapted to the high altitude, having bigger lungs, thick fur and underdeveloped sweat glands. The number of red blood corpuscles per mm^3 in sheep increases with altitude.

The Chinese have introduced finer haired sheep into the area from Mongolia and these have been cross-bred with the Tibetan sheep. The result is that better quality wool is now being produced that is suitable for spinning. Cattle from the Hulun Bair grassland of Inner Mongolia have also been introduced and cross-bred with the local cattle. The result is a cow that is 10 cm taller and gives between three and four times as much milk. This should help diversify the farming in the area in the future.

FORESTRY

At lower altitudes in areas of alpine permafrost, the lower limit of permafrost usually lies above treeline; the possibility of interaction of forestry and permafrost only occurs at high latitudes. Unless there are substantial settlements or suitable transportation routes to other areas, the forest is merely exploited to supply the local needs of the isolated homesteads and mines.

North America

In North America, the northern coniferous forest is dominated by black spruce (*Picea mariana*), which makes up 80% of the trees. It is particularly abundant near the northern forest limit at poorly drained sites. Until a commercial use has been found for it, this species is essentially unused. The logs are too thin and spindly for making log cabins but they can be used for pulpwood. However, the latter is not needed in the north. The logs can also be used for pit-props.

At better drained sites, the white spruce (*Picea glauca*) may be found together with trembling aspen (*Populus tremuloides*) and balsam poplar (*Populus balsamifera*). The white spruce provides the logs for cabins, while the wood of the poplars is suitable for firewood. Further south, the tamarack (*Larix laricina*), lodgepole pine (*Pinus contorta* var. *latifolia*) and jack pine (*Pinus banksiana*) are present and supply utility poles, railway ties, posts and mine timbers.

In most permafrost areas, the lumber industry is of limited importance, the wood being used to supply local needs only. However, the forests remain an integral part of subsistence (Alaska Geographic, 1985).

USSR

The permafrost zones of eastern Siberia extend to much lower latitudes, while treeline is usually within 400 km of the Arctic Ocean (Fig. 10.3). Furthermore, much of Siberia consists of hills and hilly plateaux which provide reasonable drainage and deeper thawing of the active layer. As a result, the forest contains good usable timber. About 40% of the forest with the highest density of canopy occurs in the permafrost areas, while about 45% of the Russian forests are situated on permafrost but a long way from the industrial areas.

At present, intensive logging is confined to the margins of the populous agricultural and industrial zones, especially in European Russia (Barr, 1979). The next zone schedule for logging includes the forests of southwest central Siberia (Fig. 10.3, after USSR, 1980). These areas include substantial areas of discontinuous permafrost where the deforestation may cause the development of thermokarst. Due to the presence of considerable quantities of ice (much of it relict from the colder Pleistocene events), this may produce substantial environmental damage and alteration. Environmental

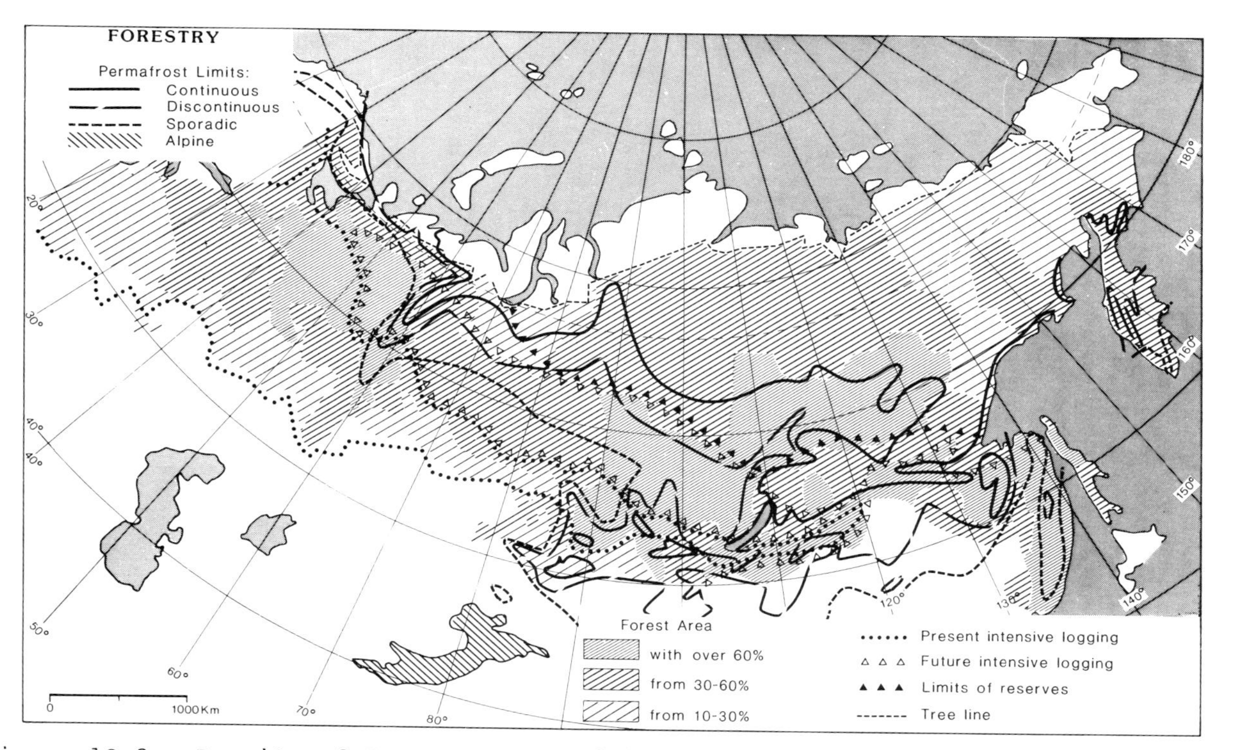

Figure 10.3: Density of Forest Cover, Limits of Present and Future Intensive Logging, and Limits of Forest Reserves in Siberia (after USSR, 1980).

damage would be expected to be less in the continuous permafrost zone where the ground is colder, although the presence of widespread alas valleys near Yakutsk (Solov'yev, 1962, 1973a, 1973b) suggests that some damage by melting of ground ice will occur in most areas.

REFERENCES

Aamot, H.W.C. (1966) Dynamic Foundation Measurements, Barter Island, Alaska. U.S. Army Cold Regions Research & Engineering Laboratory, Special Report 75. 32 pp.

Adam, K.M. (1978) Building and Operating Winter Roads in Canada and Alaska. Department of Indian & Northern Affairs, Environment Division. Environmental Studies #4. 221 pp.

Adam, K.M. and H. Hernandez (1977) 'Snow and Ice Roads: Ability to Support Traffic and Effects on Vegetation', Arctic, 30, 13-27.

Aitken, G.W. (1970) 'Transport of Frozen Soil', Proc. Vermont Conf. on Winter Construction (1969), Univ. Vermont, 50-68.

Åkerman, J.J. (1980) Studies on Periglacial Geomorphology in West Spitsbergen. Meddelanden från Lunds Univ. Geografiska Inst. Ser. Ash. 89.

Åkerman, J. (1982) 'Studies on Naledi (icings) in West Spitsbergen', Proc. Fourth Can. Permafrost Conf., Calgary. National Research Council, Ottawa, pp. 189-202.

Alaska Geographic (1982) 'Alaska's Oil/Gas and Minerals Industry', Alaska Geographic, 9(4), 216 pp.

Alaska Geographic (1984) 'Alaska's Farms and Gardens', Alaska Geographic, 11(2), 142 pp.

Alaska Geographic (1985) 'Alaska's Forest Resources', Alaska Geographic, 12(2), 199 pp.

Allen, L.T. (1977) The Trans-Alaska Pipeline. Alyeska Pipeline Service Co., 2 vols.

Allen, T.B. (1985) 'Time Catches up with Mongolia', National Geographic, 167, 242-269.

Alyeska Pipeline Service Company (1977) Summary Project Description of the Trans-Alaska Pipeline System, Alyeska Pipeline Service Company, 20 pp.

Andersland, O.B. and M.R.B. Alwahhab (1983) 'Lug Behavior for Model Steel Piles in Frozen Sand', in Permafrost: Fourth Int. Conf. Proc., National Academy Press, Washington. 1524 pp, pp. 16-21.

Anderson, D.M. (1971) Remote Analysis of Planetary Water. United States Army Corps of Engineers, Cold Regions Research and Engineering Laboratory Special Report No. 154, 13 pp.

Anderson, D.M., E.J. Chamberlin, G.L. Guymon, D.L. Kane, B.D. Kay, J.R. Mackay, K. O'Neill, S.I. Outcalt and P.J.Williams (1984) Ice Segregation and Frost Heaving, National Academy Press, Washington, C.D., 72 pp.

Anisimova, N.P., N.M. Nikitina, V.M. Piguzova and V.V. Shepelyev (1973) 'Water Resources of Central Yakutia. Guidebook, Second Int. Permafrost Conf., Yakutsk, USSR, 47 pp.

Anonymous (1956) 'Permafrost Research in Northern Canada', Nature, 178, 716-717.

References

Anonymous (1982) 'Hydrocarbon Development in the Beaufort Sea-Mackenzie Delta Region', Environmental Impact Statement, vol. 2, Development Systems, 718 pp.

Armstrong, B.C., D.W. Smith and J.J. Cameron (1981) 'Water Requirements and Conservation Alternatives for Northern Communities', in D.W. Smith and S.E. Hudrey (eds.), Design of Water and Wastewater Services for Cold Climate Communities, Pergamon Press, Oxford, pp. 65-93.

Associated Engineering Services, Ltd. (1975) Department of Environment Sewage Disposal Study, Jasper Townsite. Prepared for the Environmental Protection Service, Edmonton, Alberta.

Auld, R.G., R.J. Robbins, L.W. Rosenegger and R.H.B. Sangster (1978) Pad Foundation Design and Performance of Surface Facilities in the Mackenzie Delta. Proc. Third Int. Conf. Permafrost, Edmonton. National Research Council, Ottawa, 1, pp. 765-771.

Babb, A.L., D.M. Chow, K.L. Garlid, R.P. Popovich and E.M. Woodruff (1971) The Thermo Tube, a Natural Convection Heat Transfer Device for Stabilization of Arctic Soils in Oil Producing Regions. 46th Ann. Meeting, New Orleans, Soc. Pet. Eng., Amer. Inst. Mech. Eng., Paper SPE 3618. 12 pp.

Bakakin, V.P. (1978) Basic Areas of Geocryological Research in Mining. Proc. Second Int. Conf. on Permafrost, Yakutsk. USSR Contribution. Nat. Acad. Sciences, Washington, pp. 585-586.

Baker, P.E. (1974) Experiments on Hydrocarbon Gas Hydrates in Unconsolidated Sand, in Natural Gases in Marine Sediments. I.R. Kaplan (ed.). Plenum Press, New York, pp. 227-234.

Balmér, P. (1981) Swedish Experiences with Wastewater Treatment with Special Reference to Cold Climate, in D.W. Smith and S.E. Hrudey (eds.), Design of Water and Wastewater Services for Cold Climate Communities. Pergamon Press, Oxford, pp. 125-135.

Balobayev, V.T. (1978) Rekenstruktsiya paleoclimata po sovremennym geotermicheskim dannym - Reconstruction of Paleoclimate from Present-Day Geothermal Data. Proc. Third Int. Conf. Permafrost, Edmonton. National Research Council, Ottawa, 1, pp. 10-14.

Baranov, I. Ya. (1959) Geograficheskoye rasprostraneniye sezonne-promerzayuschchikh pochv i mnogoletnemerzlykh gornyk porod. Inst. Merzlotovedeniya in V. A. Obrucheva. Osnovy geokriologii (merzlotovedeniya), Chast' pervaya, Obshchaya geokriologiya. Moskva. Akad. Nauk S.S.S.R. 459 pp.

Baranov, I. Ya. (1964) Geographical Distribution of Seasonally Frozen Ground and Permafrost. National Research Council of Canada, Technical Translation #1121, 85 pp.

Barr, B.M. (1979) Soviet Timber: Regional Supply and Demand, 1970-1990, Arctic, 32, 308-328.

Barsch, Van D. (1977) Eine Abschätzung von Schuttproduktion und Schutttransport im Bereich aktiver Blockgletscher der Schweizer Alpen. Zeitschrift für Geomorphologie, 21, 79-86.

Bauer, A., G.R. Harris, L. Lang, P. Prezioni and D.J. Selleck (1965) How 10¢ Puts Crater Research to Work, Engineering & Mining, 166(9), 117-121.

Bauer, A., P.N. Calder, R.R. Maclachlan and M. Halupka (1973) Cratering and Ditching with Explosives in Frozen Soils. Defense Research Board, Canada. Report DREV R-699/73, 123 pp.

Baulin, V.V. (1962) Osnovnye etapy istorii razvitiya mnogoktnemerzlylch porod na territorii Zapadno-Sibirskoy nizmennosti. Akademiya Nauk SSSR. Institut Merzlotovedeniya in V.A. Obrucheva Trudy 19, 5-18.

Beauchamp, J.C. and S. Stamer (1971) Northern Bridge Construction. Proc. Roads & Transportation Assoc. Can. Convention, Vancouver, pp. 83-96.

References

Beistline, E.H. (1966) Placer Mining in Frozen Ground. Proc. First Int. Conf. Permafrost (1963), Lafayette, Indiana. Nat. Acad. Sciences, Washington. Publication 1287, pp. 463-67.

Belopukhova, E.B. (1973) Osobennasti sovremennogo razvitiya mnogoletnemerzlykh porod zapandnoy Sibiri, in Akadamiya Nauk SSSR, Sektsiya Nauk o Zemla, Sibirskoye Otdeleniye, 11 Mazhdunarodnaya Konferentsiya po Merzlotovedeniyu, Doklady: Soobshcheniya 2. 154 pp, pp. 84-86.

Benedict, J.B. (1970) Downslope Soil Movement in a Colorado Alpine Region: Rates, Processes, and Climatic Significance. Arctic & Alpine Research, 2, 165-226.

Berg, R.L. (1974) Design of Civil Airfield Pavements for Seasonal Frost and Permafrost Conditions. U.S. Federal Aviation Agency, Research & Development, Report FAA-AD-74-30. 98 pp.

Berg, R.L. (1984) Status of Numerical Models for Heat and Mass Transfer in Frost-susceptible Soils. Permafrost. 4th Int. Conf. Fairbanks Final Proc. National Academy Press, Washington, pp. 67-71.

Berg, R.L. and G.W. Aitken (1973) Some Passive Methods of Controlling Geocryological Conditions in Roadway Construction. Proc. Second Int. Conf. Permafrost, Yakutsk, USSR. North American Contribution, Nat. Acad. Sciences, Washington, pp. 581-586.

Berg, R.L. and D.C. Esch (1983) Effect of Color and Texture on the Surface Temperature of Asphalt Concrete Pavements. Permafrost. 4th Int. Conf. Proc. Fairbanks. National Academy Press, Washington. 1524 pp. pp. 57-61.

Berg, R.L. and W.F. Quinn (1977) Use of Light-Coloured Surface to Reduce Seasonal Thaw Penetration Beneath Embankments on Permafrost. Proc. Second Int. Symposium on Cold Regions Engineering (1976), Fairbanks, Alaska. Univ. Alaska, Dept. of Civil Engineering, pp. 86-99.

Berg, R.L. and N. Smith (1976) Observations Along the Pipeline Haul Road Between Livengood and the Yukon River. U.S. Army, Cold Regions Research & Engineering Lab. Special Report 76-11. 83 pp.

Bily, C. and J.W.L. Dick (1974) Naturally Occurring Gas Hydrates in the Mackenzie Delta, N.W.T. Bull. Can. Pet. Geol., 22, 340-352.

Biyanov, G.F. (1965) Construction of a Water Storage Dam on Permafrost. National Research Council, Ottawa, Technical Translation 1353. 30 pp.

Biyanov, G.F. (1975) Dams on Permafrost. U.S. Army, Cold Regions Research & Engineering Lab. Draft Translation TL 555. 234 pp.

Biyanov, G.F. and V.I. Makarov. (1980) Characteristics of the Construction of Frozen Dams in Western Yakutiya. English Translations, Part II. Proc. Third Int. Conf. Permafrost, National Research Council, Ottawa, pp. 213-226.

Bogoslovskiy, P.A., V.A. Veselov, S.B. Ukhov, A.V. Stotsenko and A.A. Tsvid (1966) Dams in Permafrost Regions. Proc. 1st Int. Conf. Permafrost (1963), Lafayette, Indiana. Nat. Acad. Sciences, Washington, Publication 1287, pp. 450-455.

Black, R.R. (1963) Les coins de glace et le gel permanente dans le nord de L'Alaska. Annales de Géographie, 72, 257-271.

Black, R.F. (1974) Ice-wedge Polygons of Northern Alaska. D.R.Coates (ed.), "Glacial Geomorphology", Fifth Annual Geomorphology Series, Binghampton, New York. 398 pp. pp. 247-275.

Blasco, S.M. (1984) A Perspective on the Distribution of Subsea Permafrost on the Canadian Beaufort Continental Shelf. Permafrost. 4th Int. Conf., Fairbanks. Final Proc. National Academy Press, Washington, pp. 83-86.

Bliss, L.C. (1962) Adaptations of Arctic and Alpine Plants to Environmental Conditions. Arctic, 15, 117-144.

References

Bliss, L.C. (1978) Polar Climates: Their Present Agricultural Uses and Their Estimated Potential Production in Relation to Soils and Climate. Plenary Session Papers. Eleventh Congress, Int. Soc. Soil Science, Edmonton, 2, 70-90.

Bliss, L.C. and R.W. Wein. (1972) Botanical Studies of Natural and Man Modified Habitats in the Eastern Mackenzie Delta Region and the Arctic Islands. Dept. Indian and Northern Affairs, Ottawa. ALUR. 71-72-14.

Bolotin, N. (1980) Klondike Lost. Alaska Geographic 7(4), 128.

Bowley, W.W. and M.D. Burghardt. Thermodynamics and Stones. EOS, Trans. Amer. Geophysical Union, 52, 407.

Boyev, V.R. and V.M. Gabov (1981) Sel'skoye khozyaystvo v rayonakh promyshlennogo osvoyeniya Sibiri. Moscow.

Brennan, A. (1982) Permafrost: A Bibliography, 1978-1982. World Data Center for Glaciology. Cooperative Institute for Research in the Environmental Sciences. Report GD14. 1-162; 2-172.

Brewer, M. (1958) Some Results of Geothermal Investigations of Permafrost in Northern Alaska. Trans. Amer. Geophysical Union, 39, 19-26.

Brewer, M.C. (1984) Petroleum Exploration and Protection of the Environment on the National Petroleum Reserve in Alaska. Permafrost. 4th Int. Conf., Fairbanks. Final Proc. National Academy Press, Washington, pp. 133-134.

Brewer, R. and A.D. Haldane (1957) Preliminary Experiments in the Development of Clay Orientation in Soils. Soil Science, 84, 301-307.

Bridgeman, P.W. (1912) Water in the Liquid and Fine Solid Forms, Under Pressure. Proc. Amer. Acad. Arts & Sciences, 47, 441-558.

Brown, Jerry and N.A. Grave (1979a) Physical and Thermal Disturbance and Protection of Permafrost. Proc. 3rd Int. Conf. Permafrost, Edmonton. National Research Council, Ottawa, 2, 51-91.

Brown, J. and N.A. Grave (1979b) Physical and Thermal Disturbance and Protection of Permafrost. U.S. Army Corps of Engineers, Cold Regions Research and Engineering Lab. Spec. Rept. 79-5. 42 pp.

Brown, J., W. Rickard and D. Vietor (1969) The Effect of Disturbance on Permafrost Terrain. U.S. Army, Cold Regions Research and Engineering Lab., Special Rept. 138. 13 pp.

Brown, J. and P.V. Sellman (1973) Permafrost and Coastal Plain History of Arctic Alaska, in M.E. Britton (ed.), Alaskan Arctic Tundra. Arctic Institute of North America, Technical Paper 25. 244 pp.

Brown, J. and Yin-Chao Yan (1982) The Second National Chinese Conference on Permafrost. U.S. Army, Cold Regions Research and Engineering Lab., Special Report 82-3. 58 pp.

Brown, R.J.E. (1960) The Distribution of Permafrost and its Relation to Air Temperature in Canada and the USSR. Arctic, 13, 163-177.

Brown, R.J.E. (1963) Proc. First Can. Conf. Permafrost. National Research Council, Assoc. Committee on Soil and Snow Mech., Ottawa. Tech. Mem. 76. 231 pp.

Brown, R.J.E. (1965) Factors Influencing Discontinuous Permafrost in Canada. Abstracts Int. Assoc. Quat. Res, 7th Int. Cong., Boulder, Col., p. 47.

Brown, R.J.E. (1966a) Influence of Vegetation on Permafrost, in Permafrost: Int. Conf., Lafayette. Nat. Acad. Sciences. Washington, Publication 1287, pp. 20-25.

Brown, R.J.E. (1966b) The Relationship Between Mean Annual Air and Ground Temperatures in the Permafrost Regions of Canada. Proc. 1st Int. Permafrost Conf. Nat. Acad. Sciences, Washington, Publication 1287, pp. 241-246.

Brown, R.J.E. (1967) Permafrost Map of Canada. National Research Council, Div. Building Res., Ottawa, NRCC 9769. Geol. Sur. Can. Map 1246A.

References

Brown, R.J.E. (1968) Occurrence of Permafrost in Canadian Peatlands. Proc. 3rd Int. Peat Cong., Quebec, pp. 174-181.

Brown, R.J.E. (1969) Factors Influencing Discontinuous Permafrost in Canada, in T.L. Péwé (ed.), "The Periglacial Environment, Past and Present." INQUA. 7th Cong., Alaska. McGill-Queens Univ. Press, Montreal, pp. 11-53.

Brown, R.J.E. (1970a) Permafrost in Canada - its Influence on Northern Development. Univ. of Toronto Press. 234 pp.

Brown, R.J.E. (1970b) Permafrost as an Ecological Factor in the Subarctic. Symp. Ecol. of Subarctic Regions, Helsinki, 1966, pp. 129-140.

Brown, R.J.E (1972) Permafrost in the Canadian Arctic Archipelago. Zeitschrift für Geomorphologie, 13, 102-130.

Brown, R.J.E. (ed.) (1973a) Proc. 1st Can. Conf. on Permafrost. National Research Council, Assoc. Committee on Soil and Snow Mech., Ottawa. Tech. Mem. 76. 231 pp.

Brown, R.J.E. (1973b) Influence of Climatic and Terrain Factors on Ground Temperature at Three Locations in the Permafrost Region of Canada. North American Contribution, Permafrost. 2nd Int. Conf., Yakutsk. Nat. Acad. Sciences, Washington, pp. 27-34.

Brown, R.J.E. (1974) Some Aspects of Airphoto Interpretation of Permafrost in Canada. Division of Building Res., National Research Council, Ottawa, Technical Paper 409. 20 pp.

Brown, R.J.E. (1978) Permafrost. Plate 32 in "Hydrological Atlas of Canada". Ottawa, Fisheries & Environment, Canada. 34 plates.

Brown, R.J.E. and G.H. Johnston (1964) Permafrost and Related Engineering Problems. Endeavour, 23, 66-72.

Brown, R.J.E. and T.L. Péwé (1973) Distribution of Permafrost in North America and its Relationship to the Environment: A Review, 1963-1973. North American Contribution, Permafrost. 2nd Int. Conf., Yakutsk. Nat. Acad. Sciences, Washington, pp. 71-100.

Bundtzen, T.K. (1982) Alaska's Strategic Minerals. Alaska Geographic, 9(4), 52-63.

Burns, B.M. (1973) The Climate of the Mackenzie Valley-Beaufort Sea. Vol. 1. Dept. of the Environment, Atmospheric Environment Service, Climatological Studies 24. 227 pp.

Cameron, J.J. (1977) Community Water Use Summary. Internal Report, Northern Technology Unit, Environmental Protection Service, Edmonton, Alberta.

Cameron, J.J., V. Christensen and D.J. Gamble (1977) Water and Sanitation in the Northwest Territories: An Overview of the Setting, Policies and Technology. The Northern Engineer, 9(4), 4-12.

Cameron, J.J. and D.W. Smith (1977) Annotated Bibliography on Northern Environmental Engineering, 1974-75. Environment Canada, Environmental Protection Service, Report EPS-3-WP-77-6. 154 pp.

Carefoot, E.I., A.L. Davies, G.H. Johnston, N.A. Lawrence, P. Lukomskyj and D.E. Thornton (1981) Utilities, in G.H. Johnston (ed.), Permafrost Engineering Design and Construction. Chapter 10. J. Wiley & Sons, Toronto, pp. 415-472.

Carey, K.L. (1973) Icings Developed from Surface Water and Ground Water. U.S. Army Cold Regions Research and Engineering Lab. Science and Engineering Monograph III-D3. 65 pp.

Carey, K.L., R.W. Huck and D.A. Gaskin (1975) Prevention and Control of Culvert Icing. U.S. Army, Cold Regions Research & Engineering Lab. Special Rept. SR 224.

Casagrande, A. (1932) A New Theory of Frost Heaving: Discussion. Proc. U.S. Highway Res. Board, 11(1), 168-172.

Chaban, P.D. and V.G. Gol'dtman (1978) Effect of Geocryological Conditions on the Productivity and Mining Operations in the Northeast USSR. Proc. 2nd Int. Conf. Permafrost, Yatutsk. USSR Contribution, Nat. Acad. Sciences, Washington, pp. 587-590.

References

Chamberlain, E.J. (1983) Frost Heave of Saline Soils, in Permafrost: 4th Int. Conf. Proc., National Academy Press, Washington. 1525 pp. pp. 121-126.

Chambers, M.J.G. (1967) Investigations of Patterned Ground at Signy Island, South Orkney Islands. III. Miniature Patterns, Frost Heaving and General Conclusions. Bull. British Antarctic Sur., 12, 1-22.

Charles, J.L. (1959) Permafrost Aspects of the Hudson Bay Railroad. J. Soil Mech. & Foundation Div., Proc. Amer. Soc. Civil Eng., 85(SM6), 125-135.

Charles, J.L. (1965) The Great Slave Lake Railway. J. Eng. Inst. Can, May, 15-19.

Chan, X., Y. Wang and J. Ping (1983) Influence of Penetration Rate, Surcharge Stress, and Ground Water Table on Frost Heave, in Permafrost: 4th Int. Conf. Proc. National Academy Press, Washington. 1524 pp. pp. 131-135.

Chan, Xiaobai (1984) Current Developments in China on Frost-Heave Processes in Soil. Permafrost. 4th Int. Conf., Fairbanks. Final Proc. National Academy Press, Washington, pp. 55-60.

Cheng Guodong (1982) The Forming Process of Thick Layered Ground Ice. Scientia Simica (Series B), 25, 777-788.

Cheng Guodong (1983) Three-Dimensional Zonation of High Latitude Permafrost. Permafrost. 4th Int. Conf. Proc., Fairbanks. National Academy Press, Washington, pp. 136-141.

Cheng Guodong (1984) Study of Climate Change in the Permafrost Regions of China - a Review. Permafrost. 4th Int. Conf. Fairbanks. Final Proc. National Academy Press, Washington, pp. 139-144.

Cherskii, N.V. and E.A. Bondarev (1972) Thermal Methods of Exploiting Gas-Hydrated Strata. Soviet Physics-Doklady, 17, 211-213.

Cherskii, N. and Yu. Makogan (1970) Solid Gas - World Reserves are Enormous. Oil & Gas International, 10(8), 82-84.

Chou, Yu-Wu and Y.-H. Tu (1963) Ching tsang kao yüan tung t'u ch'u pu k'ao ch'a. Kexue Tongbao (K'o Ksüeh T'ung Pao), 25, 60-63.

Church, M. (1974) Hydrology and Permafrost with Reference to Northern North America, in Permafrost Hydrology. Proc. Workshop Seminar, 1974. Can. Nat. Committee, the Int. Hyd. Decade, pp. 7-20.

Clark, E.F. and O.W. Simoni (1976) A Survey of Road Construction and Maintenance Problems in Central Alaska. U.S. Army, Cold Regions Res. and Engineering Lab., Special Rept. SR 76-8. 36 pp.

Collett, T.S. (1983) Detection and Evaluation of Natural Gas Hydrates from Well Logs, Prudhoe Bay, Alaska, in Permafrost: 4th Int. Conf. Proc. Fairbanks. National Academy Press, Washington. 1524 pp. pp. 169-174.

Cook, D.J. (1983) Placer Mining in Alaska. School of Mineral Industry, University of Alaska, Fairbanks.

Cook, F.A. (1966) Patterned Ground Research in Canada. Permafrost. 1st Int. Conf., Lafayette, Indiana. Proc. Nat. Acad. Sciences, Washington. Publication 1287, pp. 128-130.

Cooper, P.F. (1975) Movement and Deformation of the Landfast Ice of the Southern Beaufort Sea. Beaufort Sea Technical Rept. 37. Victoria, Dept. of the Environment. 16 pp.

Corbel, J. (1956) Phénomènes périglaciaires au Svalbard et en Laponie. Rapports de la Commission de morphologie périglaciaire de l'I.G.U. Biuletyn Peryglacjalny, 4, 47-54.

Corte, A.E. (1961) The Frost Behavior of Soils: Laboratory and Field Data for a New Concept - I. Vertical Sorting. U.S. Army Corps of Engineering, Cold Regions Res. and Engineering Lab. Research Rept. 85(1). 22 pp.

Corte, A.E. (1962a) Vertical Migration of Particles in Front of a Moving Freezing Plane. J. Geophysical Res., 67, 1085-1090.

Corte, A.E. (1962b) Relationship Between Four Ground Patterns, Structure of the Active Layer, and Type and Distribution of Ice in the Permafrost. U.S. Army Corps of Engineers, Cold Regions Res. and Engineering Lab. Research Rept. 88. 79 pp.

Corte, A.E. (1962c) The Frost Behavior of Soils: Laboratory and Field Data for a New Concept - II, Horizontal Sorting. U.S. Army Corps of Engineers, Cold Regions Res. and Engineering Lab. Research Rept. 85(2). 20 pp.

Corte, A.E. (1962d) The Frost Behavior of Soils - I, Vertical Sorting, in Soil Behavior on Freezing With and Without Additives. Nat. Acad. Sciences, Washington. N.R.C. Highway Res. Bd. Bull. 317. 34 pp. pp. 9-34.

Corte, A.E. (1962e) The Frost Behavior of Soils - II, Horizontal Sorting, in Soil Behavior Associated With Freezing. Nat. Acad. Sciences,Washington. N.R.C. Highway Res. Bd. Bull 331. 115 pp. pp. 44-66.

Corte, A.E. (1963a) Vertical Migration of Particles in Front of a Moving Freezing Plane. U.S. Army Cold Regions Res. and Engineering Rept. 105. 8 pp.

Corte, A.E. (1963b) Particle Sorting by Repeated Freezing and Thawing. Science, 142, 499-501.

Corte, A.E. (1966a) Experiments on Sorting Processes and the Origin of Patterned Ground, in Permafrost: 1st Int. Conf. Proc., Lafayette. Indiana. Nat. Acad. Sciences, Washington. Publication 1287. 563 pp. pp. 130-135.

Corte, A.E. (1966b) Particle Sorting by Repeated Freezing and Thawing. Biuletyn Peryglacjalny, 15, 175-240.

Croasdale, K.R. and R.W. Marcellus (1978) Ice and Wave Action on Artificial Islands in the Beaufort Sea. Can. J. Civil Eng., 5, 98-113.

Croasdale, K.R., M. Metge and P.H. Verity (1978) Factors Governing Ice Ride-up on Sloping Beaches, in Int. Assoc. Hydraulic Res. Symp. Ice Problems, Lulea, Sweden. Proc. Part I, Ice Forces on Structures, 525 pp. pp. 405-420.

Cronin, J.E. (1983) Design and Performance of a Liquid Natural Convective Subgrade Cooling System for Construction on Ice-Rich Permafrost, in Permafrost: 4th Int. Conf. Proc. National Academy Press, Washington. 1524 pp. pp. 198-203.

Crory, F.E. (1965) Pile Foundations in Discontinuous Permafrost Areas. Proc. Can. Regional Permafrost Conf., Edmonton. N.R.C. Assoc. Committee Soil and Snow Mech. Technical Mem. 86, pp. 58-76.

Crory, F.E. (1966) Pile Foundations in Permafrost. Proc. 1st Int. Conf. Permafrost, Lafayette. Nat. Acad. Sciences, Washington. Publication 1287, pp. 467-476.

Crory, F.E. (1968) Bridge Foundations in Permafrost Areas: Goldstream Creek, Fairbanks, Alaska. U.S. Army, Cold Regions Res. and Engineering Lab., Technical Rept. 180. 33 pp.

Crory, F.E. (1973) Installation of Driven Test Piles in Permafrost at Bethel Air Force Station, Alaska. U.S. Army, Cold Regions Res. and Egnineering Lab., Technical Rept. 145. 27 pp.

Crory, F.E. (1975) Bridge Foundations in Permafrost: Moose and Spinach Creeks, Fairbanks, Alaska. U.S. Army, Cold Regions Res. & Engineering Lab., Technical Rept. 266, 36 pp.

Crory, F.E. (1978) The Kotzebue Hospital: A Case Study. Proc. Conf. Applied Techniques for Cold Environments, Anchorage, Alaska. Amer. Soc. Civil Eng., 1, 342-359.

Crory, F.E. (1982) Piling in Frozen Ground. J. of Tech. Councils, Amer. Soc. Civil Eng., 108(TC1), 112-124.

Crory, F.E. (1985) Long-Term Foundation Studies of Three Bridges in the Fairbanks Area. U.S. Army Cold Regions Res. and Engineering Lab., Technical Rept.

Crowley, V.F. (1977) Material Selection and Design Practice for Electric Power Stations in the Arctic. Proc. Int. Conf. Materials Engineering in the Arctic (1976), St. Jovite, Quebec. M.B. Ives (ed.). Amer. Soc. Metals, pp. 71-77.

References

Cui Zhijiu (1980) Periglacial Phenomena and Environmental Reconstructions on the Qinghai-Xizang Plateau, in Scientific Papers on Geology for International Exchange 5. Beijing. Publishing House of Geology, pp. 109-117.

Cui Zhijiu (1983) An Investigation of Rock Glaciers in the Kuulun Shan, China. Permafrost. 4th Int. Conf., Fairbanks. Proc. National Academy Press, Washington, pp. 208-211.

Czeppe, Z. (1959) Uwagi o procesie wymarzania glazow. Czasopismo Geograficzne, 30, 195-202.

Czeppe, Z. (1960) Thermic Differentiation of the Active Layer and its Influence Upon the Frost Heave in Periglacial Regions (Spitsbergen). Bull. Acad. polonaise Sciences, Séries Science géolog. et géogr., 8, 149-152.

Czeppe, Z. (1966) Przebieg glównych procesow morfogenetycznych w potudniowozachadnim Spitsbergenie. Uniwersytetu Jagiellońskiego Zeszyty Naukowe Prace Geograficzne, Zeszyty 13, Prace Instytutu Geograficzneyo Zeszylt, 35. 129 pp.

Czudek, T. and J. Demek (1970) Thermokarst in Siberia and its Influence on the Development of Lowland Relief. Quat. Res., 1, 103-120.

Darwin, D. (1832) The Voyage of the Beagle. Bantam Classic, 1958, p. 213.

Davison, D.M. and R.C. Lo (1982) Preservation of Permafrost for a Fuel Storage Tank. Proc. 4th Can. Permafrost Conf., Calgary. N.R.C., Ottawa. 591 pp. pp. 545-554.

Davidson, D.W. (1973) Clathrate Hydrates, in F. Frank (ed.), Water: A Comprehensive Treatise. Plenum Press, New York, vol. 2, pp. 115-234.

Davidson, D.W., K.K. El-Defrawy, M.O. Fuglem and A.S. Judge (1978) Natural Gas Hydrates in Northern Canada. Proc. 3rd Int. Conf. Permafrost, Edmonton. N.R.C., Ottawa, vol. 1, pp. 937-943.

Day, J.H., and H.M. Rice. (1964) The Characteristics of Some Permafrost Soils in the Mackenzie Valley, N.W.T. Arctic, 4, 222-236.

Dement'ev, A.I. (1959) Principles of Geocryology, Part II. N.R.C., Ottawa, Technical Translation 1287.

Department of Agriculture (1970) Gardening on Permafrost. Ottawa, Publication 1408.

Department of Agriculture (1977) Northern Agriculture. Ottawa, Publication 1616.

Dickens, H.B. and D.M. Gray (1960) Experience with a Pier-Supported Building Over Permafrost. Amer. Soc. Civil Eng., J. of Soil Mech. and Found. Div., 86(SM5), 1-14.

Ding Jinkang (1984) Study and Practice on Deep Foundation in Permafrost Areas of China. Permafrost. 4th Int. Conf., Fairbanks. Final Proc. National Academy Press, Washington, pp. 18-24.

Dingle, P.J. (1981) Island Construction in the Beaufort Sea. Esso Resources, Canada Ltd. 13 pp.

Dodolina, V.T. (1978) Wastewater Classification by Fertilizing Value. U.S. Army, Cold Regions Res. and Engineering Lab., Technical Translation TL691. 7pp.

Druzenko, A., A. Yechelev, A. Illarionov, L. Kupelyushnyy, A. Kleva, B. Reznik and V. Sukhachevskiy (1984) The Mainline and Cropland. Izvestiya, Aug. 3, 1984, p. 2. In: Soviet Geography, 25, 620-625.

Dubnie, A. (1972) Northern Mining Problems with Particular Reference to Unit Operations in Permafrost. Dept. of Energy, Mines * Resources, Ottawa. Tech. Bull. TB148. 20 pp.

Duguid, D.R., J.H. Reynolds and A.D.G. Robinson (1973) Control of Frost Heave Cracking in a Zoned Earth Dyke. Proc. Can/USSR Seminar on Civil and Mech. Eng. Aspects of the Electrical Power Industry, Moscow, 2 35 pp.

References

Dunn, J.R. and P.P. Hudec (1965) The Influence of Clays on Water and Ice in Rock Pores (2). New York State, Dept. of Public Works Physical Research Report RR65-5. 149 pp.
Dunn, J.R. and P.P. Hudec (1966) Frost Deterioration: Ice or Ordered Water? Geol. Sur. Amer. Special Paper 101. 256 pp.
Dunn, J.R. and P.P. Hudec (1972) Frost and Sorbtion Effects in Argillaceous Rocks. Highway Res. Bd: Frost Action in Soils, Nat. Acad. Sciences, Washington - Nat. Acad. Eng., Highway Res. Record, 393, 65-78.
Dutkiewicz, L. (1967 The Distribution of Periglacial Phenomena in NW-Sörkapp, Spitsbergen. Biuletin Peryglacjalny, 16,37-83.
Economist Intelligence Unit (1956) The U.S.S.R. and Eastern Europe. Oxford Univ. Press, London. 134 pp.
Eggner, C.L. and B.G. Tomlinson (1978) Temporary Wastewater Treatment in Remote Locations. J. Water Pollution Control Federation, 50, 2643-2656.
Eichfeld, I.G. (1931) Problema zemledelya na Kraynem Severe. Sovetskiy Sever, 5.
Ellenberg, L. (1974) Shimobashira-Kammeis in Japan. Geog. Helvetica, 1974, 1-5.
Emel'yanov, V.I. (1973) Measures to Prolong Overburden Removal in Permafrost Placers. (in Russian). Kolyma, 1, 6-7.
Emel'yanov, V.I. and G.Z. Perl'shtein (1980) Water Thawing of Frozen Ground for Open Pit and Underground Mining in the Northeast of the U.S.S.R. English Translations, Part II. Proc. 3rd Int. Conf. Permafrost,N.R.C., Ottawa, pp. 339-354.
Environment Canada (1982) Canadian Climate Normals, 1951-1980. Volume 2. Temperature. Atmospheric Environment Service, Ottawa. 306 pp.
Environmental Protection Service (1977) National Inventory of Municipal Waterworks and Wastewater Systems in Canada, 1975. Environmental Protection Service, Ottawa.
Esch, D.C. (1982) Thawing of Permafrost by Passive Solar Means. Proc. 4th Can. Permafrost Conf. N.R.C., Ottawa, pp. 560-569.
Esch, D.C. (1983) Evaluation of Experimental Design Features for Roadway Construction Over Permafrost. Permafrost. 4th Int. Conf., Proc., Fairbanks. National Academy Press, Washington. 1524 pp. pp. 283-288.
Esch, D.C. (1984a) Design and Performance of Road and Railway Embankments on Permafrost. Permafrost. 4th Int. Conf., Fairbanks. Final Proc. National Academy Press, Washington, pp. 25-30.
Esch, D.C. (1984b) Surface Modifications for Thawing of Permafrost. Interim Report. State of Alaska, Dept. of Transportation & Public Facilities, Rept. FHWA-AK-RD-85-10.
Essoglou, M.E. (1957) Piling Operations in Alaska. The Military Engineer, 49(330), 282-287.
Everett, D.H. (1961) The Thermodynamics of Frost Action in Porous Solids. Trans. Faraday Soc., 57, 1541-1551.
Everett, K.R. (1966) Slope Movement and Related Phenomena, in N.J. Wilimovsky (ed.), The Environment of the Cape Thompson Region, Alaska. United States Atomic Energy Commission, PNE-481, pp. 175-220.
Fahey, B.D. (1973) An Analysis of Diurnal Freeze-Thaw and Frost Heave Cycles in the Indian Peaks Region of the Colorado Front Range. Arctic and Alpine Res., 5, 269-281.
Fahey, B.D. (1974) Seasonal Frost Heave and Frost Penetration Measurements in the Indian Peaks Regions of the Colorado Frost Range. Arctic and Alpine Res., 6, 63-70.
Fedorov, N.F. and O.V. Zaborshchikov (1979) Spravochnik po proektivovanitu sistem vadosnabzheniia i kanalizatsii v raionakh vechnomerzlykh gruntov. Leningrad. Stroiizdat.
Fedorov, N.F. and O.V. Zaborshchikov (1982) Handbook on the Design of Water Supply and Sewer Systems in Regions with Permafrost Soils. U.S. Army Foreign Science and Technology Center, Technical Translation, FSTC-HT.
Fernette, G. (1982) The Mining Process. Alaska Geographic, 9(4), 120-133.

References

Ferrians, O.J. Jr. (1965a) Permafrost Map of Alaska. U.S. Geol. Sur. Miscellaneous Geologic Investigations, Map 1-445.

Ferrians, O.J. Jr. (1965b) Distribution and Character of Permafrost in the Discontinuous Zone of Alaska (Summary). Can. Regional Permafrost Conf., 1964, Proc. Tech. Mem. 86, N.R.C., Ottawa.

Ferrians, O.J., Jr. (1966) Permafrost Map of Alaska, in Proc. 1st Int. Permafrost Conf. Nat. Acad. Sciences, Washington, N.R.C., Ottawa, Publication 1287, p. 172.

Ferrians, O.J. Jr. (1984) Pipelines in the Northern U.S.S.R. Permafrost. 4th Int. Conf., Fairbanks. Final Proc. National Academy Press, Washington, pp. 98-99.

Ferrians, O.J. Jr. and G.D. Hobson (1973) Mapping and Predicting Permafrost in North America: A Review, 1963-1973. Permafrost: The North American Contribution to the 2nd Int. Conf. Permafrost. Nat. Acad. Sciences, Washington, pp. 479-498.

Ferrians, O., R. Kachadoorian and G.W. Greene (1969) Permafrost and Related Engineering Problems in Alaska. U.S. Geol. Sur., Professional Paper 678. 37 pp.

Feulner, A.J. and J.R. Williams (1967) Development of the Ground-water Supply at Cape Lisborne, Alaska, by Modification of the Thermal Regime of Permafrost. U.S. Geol. Sur. Professional Paper 575-B, pp. 199-202.

Fife, J.A. (1960) Refrigerant Piping System Supports Arctic Radar Sites. Heating, Piping and Air Conditioning, 32, 112-118.

Fletcher, R.J. (1964) The Use of Aerial Photographs for Engineering Soil Reconnaissance in Arctic Canada. Photogrammetric Engineering, 30, 210-219.

Foster-Miller Associates (1965) Final Phase I Report of an Investigation of Methods of Conveying Snow, Ice and/or Frozen Ground from an Excavation to a Disposal Area. U.S. Army, Cold Regions Res. and Eng. Lab., Hanover, New Hampshire. Internal Rept. IR 23. 101 pp.

Fotiyev, S.M., N.S. Danilova and N.S. Sheveleva (1974) Geokriologicheskiya usloviya Sredney Sibiri. Moskva, "Nauka". 147 pp.

Fraser, D.C. and P. Hoekstra (1976) Permafrost and Gravel Delineation Using Airborne Resistivity Maps from a Multicoil Electromagnetic System. Presented at 46th Ann. Meeting, Soc. Exploration Geophysicists, Houston, Texas.

Frederking, R. and B.D. Wright (1980) Characteristics of an Ice-Rubble Field at Issungnak, February-March, 1980. Proc. N.R.C. Workshop on Ridges and Rubble Fields, Calgary.

Freedman, B. (1977) Sanitarian's Handbook - Theory and Administrative Practice for Environmental Health. Peerless Publishing Co., New Orleans.

French, H.M. (1970) Soil Temperatures in the Active Layer, Beaufort Plain. Arctic, 23, 229-239.

French, H.M. (1975) Man-Induced Thermokarst, Sachs Harbour Airstrip, Banks Island, N.W.T. Can. J. Earth Sciences, 12, 132-144.

French, H.M. (1978) Sump Studies I: Terrain Disturbances. Environmental Studies No. 6, Indian and Northern Affairs, Ottawa. 52 pp.

French, H.M. (1980) Terrain, Land Use and Waste Drilling Fluid Disposal Problems, Arctic Canada. Arctic, 33, 794-806.

French, H.M. (1984) Terrain and Environmental Problems Associated with Exploratory Drilling, Northern Canada. Permafrost. 4th Int. Conf., Fairbanks. Final Proc. National Academy Press, Washington, pp. 129-132.

French, H.M., S.A. Harris and R.O. Van Everdingen (1983) The Klondike and Dawson, in H.M.French and J.A. Heginbottom (eds.) "Guidebook to Permafrost and Related Features of the Northern Yukon Territory and Mackenzie Delta, Canada." Guidebook 3. 4th Int. Conf. Permafrost, Fairbanks. Div. of Geological and Geophysical Surveys, State of Alaska. 186 pp. pp. 35-63.

References

Froehlich, W. and J. Słupik (1978 Frost Mounds as Indicators of Water Transmission Zones in the Active Layer of Permafrost During the Winter Season (Khangay Mountains, Mongolia). Proc. 3rd Int. Conf. Permafrost, Edmonton, 1, pp. 188-193.

Frost, R.E., J.H. Malerran and R.D. Leighty (1966) Photo-Interpretation in the Arctic and Subarctic. Proc. 1st Int. Permafrost Conf., Lafayette, Indiana. Nat. Acad. Sciences Publication 1287, pp. 343-348.

Fuglestad, T.C. (1985) The Alaskan Railroad Between Anchorage and Fairbanks. Guidebook 6. 4th Int. Conf. Permafrost, Fairbanks.

Fujii, Y. and K. Higuchi (1976) Ground Temperature and its Relation to Permafrost Occurrences in the Khumba Region and Hidden Valley, Nepal, Himalayas. Seppyo, 38, 125-128.

Fujii, Y. and K. Higuchi (1978) Distribution of Alpine Permafrost in the Northern Hemisphere and its Relation to Air Temperature. Proc. 3rd Int. Conf. Permafrost, Edmonton. N.R.C., Ottawa, 1, pp. 366-371.

Fullard, H. and R.F. Treharne (eds.) (1962) Muir's Historical Atlas. George Philip and Son Ltd., London. Second Edition. 96 pp.

Fulwider, C.W. (1973) Thermal Regime in an Arctic Earthfill Dam. Proc. 2nd Int. Conf. Permafrost, Yakutssk, USSR. North American Contribution. Nat. Acad. Sciences, Washington, pp. 622-628.

Fulwider, C.W. and G.W. Aitken (1962) Effect of Surface Colour on Thaw Penetration Beneath an Asphalt Surface in the Arctic. Proc. Int. Conf. on the Structural Design of Asphalt Pavements, Univ. Michigan, pp. 605-610.

Gamble, D.J. and C.T.L. Janssen (1974) Evaluating Alternative Levels of Water and Sanitation Service for Communities in the Northwest Territories. Can. J. Civil Eng., 1, 116-128.

Gandahl, R. (1979) Some Aspects on the Design of Roads with Boards of Plastic Foam. Proc. 3rd Int. Conf. Permafrost, Edmonton. N.R.C., Ottawa. 947 pp. pp. 791-797.

Garg, O.P. (1973) In situ Physicomechanical Properties of Permafrost Using Geophysical Techniques. Proc. 2nd Int. Conf. Permafrost, Yakutsk, USSR. North American Contribution. Nat. Acad. Sciences, Washington, pp. 508-517.

Garg, O.P. (1977) Applications of Geophysical Techniques in Permafrost Studies for Subarctic Mining Operations. Proc. Symp. on Permafrost Geophysics, Vancouver. Assoc. Committee on Geotechnical Res., N.R.C., Ottawa. Technical Mem. 119, pp. 60-70.

Garg, O.P. (1979) Mining of Frozen Iron Ore in Northern Quebec and Labrador. Géographie Physique et Quaternaire, 33, 369-376.

Garg, O.P. (1982) Recently Developed Blasting Techniques in Frozen Iron Ore at Schefferville, Quebec. The Roger J.E. Brown Memorial Volume, Proc. 4th Can. Permafrost Conf., Calgary, Alberta, March 1981, H.M. French, Ed., Associate Committee on Geotechnical Research, National Research Council of Canada, Ottawa, Ontario, pp. 586-591.

Gasser, G.W. (1948) Agriculture in Alaska. Arctic, 1, 75-83.

Gell, A. (1974) Some Observations on Ice in the Active Layer and in Massive Ice Bodies Tuktoyaktuk Coast, N.W.T. Geol. Sur. Can. Paper 74-1, Part A, p. 387.

Geological Survey of Canada (1983) Principal Mineral Areas of Canada. Geol. Sur. Can. Map 900A. 33rd ed.

Gerasimov, I.P. and R.P. Zimina (1968) Recent Natural Landscapes and Ancient Glaciation of the Pamir, in H.E. Wright, Jr. and W.H. Osburn (eds.), Arctic and Alpine Environments. Indiana Univ. Press, Bloomington & London. Chapter 18, pp. 267-269.

Gjessing, E.T. (1981) Water Treatment Considerations - Aquatic Humus, in D.W. Smith and S.E. Hrudey (eds.), Design of Water and Wastewater Services for Cold Climate Communities, pp. 95-101.

References

Gladwell, R.W. (1976) Field Studies of Eight First Year Sea Ice Pressure Ridges in the Southern Beaufort Sea. Arctic Petroleum Operators Ass. project 75, Esso Resources Rept. IRPT-8ME-76. 101 pp. Artus Micrographics Enterprises Ltd., Calgary. 101 pp.

Gold, L.W. (1975) The Ice Factor in Frozen Ground, in M. Church & O. Slaymaker (eds.), Field and Theory. Lectures in Geocryology. Univ. of British Columbia Press, pp. 74-95.

Gold, L.W. and A.H. Lachenbruch (1973) Thermal Conditions in Permafrost - a Review of North American Literature, in Permafrost: 2nd Int. Conf., Yakutsk, USSR. North American Contribution. Nat. Acad. Sciences, Washington, D.C. 783 pp. pp. 3-23.

Goodman, M.A. (1977) How Permafrost Affects Offshore Wells and Structures. World Oil, 185 (Oct.), 90-95.

Goodman, M.A. (1978a) Designing Casings and Wellheads for Arctic Service. World Oil, 186 (Feb.), 44-60.

Goodman, M.A. (1978b) Completion Equipment Fulfills Special Arctic Needs. World Oil, 186 (Mch.), 60-70.

Goodman, M.A. (1978c) Reducing Permafrost Thaw Around Arctic Wellbores. World Oil, 186 (Apr.), 71-76.

Goodman, M.A., F.J. Fischer and D.C. Garrett (1982). Thaw Subsidence Analysis for Multiple Wells on a Gravel Island. The Roger J.E. Brown Memorial Volume, Proc. 4th Can. Permafrost Conf., Calgary, Alberta, March 1981, H.M. French, Ed., Associate Committee on Geotechnical Research, National Research Council of Canada, Ottawa, Ontario, 591 pp., pp. 497-506.

Goodman, M.A. and L.J. Franklin (1982) Thermal Model of a New Concept for Hydrate Control During Drilling. Proc. 4th Can. Permafrost Conf., Calgary. N.R.C., Ottawa. 591 pp. pp. 349-355.

Goodman, M.A. and R.F. Mitchell (1978) Permafrost Thaw Subsidence Casing Design. J. Petroleum Tech., 20 (Mch.), 455-460.

Goodrich, L.E. (1983) Thermal Performance of a Section of the Mackenzie Highway. Permafrost. 4th Int. Conf., Proc., Fairbanks. National Academy Press, Washington. 1524 pp. pp. 353-358.

Goodwin, C.W., J. Brown and S.I. Outcalt (1984) Potential Responses of Permafrost to Climatic Warming, in J.H. McBeath (ed.), The Potential Effects of Carbon Dioxide-Induced Climatic Changes in Alaska. School of Agriculture and Land Resources Management, Univ. of Alaska-Fairbanks. Misc. Publ. 83-1, pp. 92-105.

Gradwell, M.W. (1957) Patterned Ground at a High-Country Station. New Zealand J. of Science and Tech., 38(B), 793-806.

Grave, N.A. (1968a) Merzlyye tolshchi zemli. Priroda, 1, 46-53.

Grave, N.A. (1968b) The Earth's Permafrost Beds. Canada Defense Res. Bd. Translation T499R, pp. 1-10.

Grave, N.A. (1983) A Geocryological Aspect of the Problem of Environmental Protection. Permafrost. 4th Int. Conf. Proc. National Academy Press, Washington, pp. 369-373.

Grave, N.A. (1984a) Development and Environmental Protection in the Permafrost Zone of the USSR: a Review. Permafrost. 4th Int. Conf., Fairbanks. Final Proc. National Academy Press, Washington, pp. 116-124.

Grave, N.A. (1984b) Cryogenic Processes Associated with Developments in the Permafrost Zone. Permafrost. 4th Int. Conf. Final Proc. National Academy Press, Washington, pp. 226-230.

Grawe, O.R. (1936) Ice as an Agent of Rock Weathering: A Discussion. J. of Geol., 44, 173-182.

Gray, J.T. (1973) Geomorphic Effects of Avalanche and Rock-Falls on Steep Mountain Slopes in the Central Yukon Territory, in B.D. Fahey & R.D. Thompson (eds.), "Research in Polar and Alpine Geomorphology", Geo Abstracts Ltd., Univ. of East Anglia, pp. 107-117.

References

Grechishchev, S.E. (1984) The Principles of Thermorheology of Cryogenic Soils. Permafrost. 4th Int. Conf., Fairbanks. Final Proc. National Academy Press, Washington, pp. 52-54.

Greene, D.F. (1983) Permafrost, Fire, and the Regeneration of White Spruce at Treeline Near Inuvik, N.W.T. Unpublished M.Sc. thesis, Dept. of Geography, Univ. of Calgary. 138 pp.

Greenstein, L.A. (1983) An Investigation of Midlatitude Alpine Permafrost on Niwot Ridge, Colorado, Rocky Mountains, U.S.A. Permafrost. 4th Int. Conf. Proc. Fairbanks. National Academy Press, Washington, pp. 380-383.

Gregory, J.S. (1968) Russian Land, Soviet People. Pegasus, New York. 947 pp.

Grossman, R.B. and F. J. Carlisle (1969) Fragipan Soils of the Eastern United States, in Advances in Agronomy, 21, 237-279. Academic Press, New York.

Guillien, Y. and J.-P Lautridou (1974) Conclusions des recherches de gélifraction expérimentale sur les calcaires des Charantes in Recherches de gélifraction expérimentale du Centre de Géomorphologie - IV. Nouveaux resultats sur facies calcaires. CNRS, Centre de Geomorphologie de Caen Bull. 19, 43 pp.

Gupta, R.C., R.G. Marshall and D. Badke (1973) Instrumentation for Dykes on Permafrost, Kettle Generating Station. Can. Geotechnical J., 10, 410-427.

Gysi, M. and G. Lamb (1977) An Example of Excess Urban Water Consumption. Can. J. Civil Eng., 4, 66-71.

Haag, R.W. and L.C. Bliss (1974a) Functional Effects of vegetation on the Radiant Energy Budget of Boreal Forest. Can. Geotechnical J., 11, 374-379.

Haag, R.W. and L.C. Bliss (1974b) Energy Budget Changes Following Surface Disturbance to Upland Tundra. J. Applied Ecology, 11, 355-374.

Haeberli, W. (1973) Die Basis-Temperatur der winterlichen Schnee-decke als möglicher Indikator fur die Verbreitung von Permafrost in den Alpen. Zeitschrift für Gletscherkunde und Glacialgeologie, 9, 221-227.

Haeberli, W. and G. Patzelt (1983) Permafrostkartierung im Gebiet der Hochebenkar-Blockgletscher, Obergurgl, Oetztalen Alpen. Zeitschrift für Gletscherkunde und Glacialgeologie, 18, 127-150.

Hamberg, A. (1915) Zur Kenntnis der Vorgänge in Erdboden beim Gefrieren und Auftauen sowie Bemerkungen über die erste Kristallisation des Eises in Wasser. Geol. Fören., Stockholm, Förh., 37, 583-619.

Hammerschmidt, E.G. (1934) Formation of Gas Hydrates in Natural Gas Transmission Lines. Industrial and Eng. Chem., 26, 851-855.

Hansen, B.L. and C.C. Langway, Jr. (1966) Deep Core Drilling in Ice Core Analysis at Camp Century, Greenland, 1961-1966. Antarctic J., U.S. vol. 1, 207-208.

Harden, D., P. Barnes and E. Reimnitz (1977) Distribution and Character of Naleds in Northeastern Alaska. Arctic, 30, 28-40.

Harding, R.G. (1962) Foundations Problems at Fort McPherson, N.W.T. Proc. 1st Can. Conf. Permafrost. N.R.C. Assoc. Committee for Geotechnical Res., Ottawa. Tech. Mem. 76, pp. 159-166.

Harris, S.A. (1969) The Meaning of Till Fabrics. Can. Geographer, 13, 317-337.

Harris, S.A. (1971) Preliminary Observations on Downslope Movement of Soil During the Fall in the Chinook Belt of Alberta, in "Research Methods in Pleistocene Geomorphology", 2nd Guelph Symp. on Geomorphology, pp. 275-285.

Harris, S.A. (1976) The Vermilion Pass Fire: The First Seven Years. Contract rept. for the Parks Branch, Dept. of Indian Affairs and Northern Development. 176 pp.

References

Harris, S.A. (1981a) Climatic Relationships of Permafrost Zones in Areas of Low Winter Snow-Cover. Arctic, 34, 64-70.

Harris, S.A. (1981b) Climatic Relationships of Permafrost Zones in Areas of Low Winter Snow-Cover. Biuletyn Peryglacjalny, 28, 227-240.

Harris, S.A. (1982a) Distribution of Zonal Permafrost Landforms with Freezing and Thawing Indices. Biuletyn Peryglacjalny, 29, 163-182.

Harris, S.A. (1982b) Identification of Permafrost Zones Using Selected Periglacial Landforms. The Roger J.E. Brown Memorial Volume, Proc. 4th Can. Permafrost Conf., Calgary, Alberta, March 1981, H.M. French, Ed., Associate Committee on Geo-Technical Research, National Research Council of Canada, Ottawa, Ontario, pp. 49-58.

Harris, S.A. (1983a) Comparison of the Climatic and Geomorphic Methods of Predicting Permafrost Distribution in Western Yukon Territory, in Permafrost: 4th Int. Conf. Proc. National Academy Press, Washington. 1524 pp. pp. 450-455.

Harris, S.A. (1983b) Cold Air Drainage West of Fort Nelson, British Columbia. Arctic, 35, 537-541.

Harris, S.A. (1985) Distribution and Zonation of Permafrost Along the Eastern Ranges of the Cordillera of North America. Biuletyn Peryglacjalny, 30, pp.

Harris, S.A. (1986) Permafrost Distribution, Zonation, and Stability Along the Eastern Ranges of the Cordillera of North America. Arctic, , pp.

Harris, S.A. and R.J.E. Brown (1978) Plateau Mountain: A Case Study of Alpine Permafrost in the Canadian Rocky Mountains. Proc. 3rd Int. Conf. Permafrost, Edmonton, 1, pp. 385-391.

Harris, S.A. and R.J.E. Brown (1982) Permafrost Distribution Along the Rocky Mountains in Alberta. Proc. 4th Can. Permafrost Conf., Calgary. N.R.C., Ottawa, pp. 59-67.

Harris, S.A., J.A. Heginbottom, C. Tarnocai and R.O. van Everdingen (1983) The Dempster Highway - Eagle Plain to Inuvik, in H.M. French and J.A. Heginbottom (eds.), "Guidebook to Permafrost and Related Features of the Northern Yukon Territory and Mackenzie Delta, Canada." 4th Int. Conf. Permafrost, Fairbanks. 186 pp, pp. 87-111.

Harris, S.A., H.M. French, J.A. Heginbottom, G.H. Johnston, B. Ladanyi, D.C. Sego and R.O. van Everdingen (1986) Glossary of Permafrost Related Ground Ice Terms. National Research Council of Canada, Assoc. Committee on Geotechnical Res., Permafrost Sub-Committee. Ottawa.

Harwood, T.A. (1966) Dew Line Site Selection and Explanation. Proc. 1st Int. Conf. Permafrost, Lafayette, Indiana. Nat. Acad. Sciences, Washington. Publication 1287, pp. 359-363.

Hattersley-Smith, G. (1963) The Ward Hunt Ice Shelf: Recent Changes in the Ice Front. J. of Glaciology, 4, 415-424.

Hay, T. (1936) Stone Stripes. Geographical J., 87, 47-50.

Hayley, D.W. (1982) Application of Heat Pipes to Design of Shallow Foundations on Permafrost. Proc. 4th Can. Permafrost Conf., Calgary. N.R.C., Ottawa. 591 pp. pp. 535-544.

Hayley, D.W. (1984) Geotechnical and Engineering Significance of Subsea Permafrost. Permafrost. 4th Int. Conf., Fairbanks. Final Proc. National Academy Press, Washington, pp. 93-95.

Hayley, D.W., W.D. Roggensack, W.E. Jubien and P.V. Johnson (1983) Stabilization of Sinkholes on the Hudson Bay Railway, in Permafrost: 4th Int. Conf. Proc. National Academy Press, Washington. 1524 pp. pp. 468-473.

He Changgeng (1983) Building Foundations on Permafrost, in Permafrost: 4th Int. Conf. Proc. National Academy Press, Washington. 1524 pp. pp. 474-479.

Heginbottom, J.A. ((1971) Some Effects of a Forest Fire on the Permafrost Active Layer at Inuvik, N.W.T. N.R.C., Assoc. Committee on Geotechnical Res., Tech. Mem. 103, pp. 31-36.

References

Heginbottom, J.A. (1973) Some Effects of Surface Disturbance on the Permafrost Active Layer at Inuvik, N.W.T. Environmental-Social Committee on Northern Pipelines, Rept. 73-16. 23 pp.

Heginbottom, J.A. (1984) The Mapping of Permafrost. Can. Geographer, 28, 78-83.

Heginbottom, J.A. and M. Sinclair (1985) A Cumulative Index to Permafrost Conference Proceedings, 1958-1983. Geol. Sur. Can., Open File Rept. 1135. 213 pp.

Henderson, J.D. (1980) Permafrost Mapping Along Transportation Corridors. Proc. Symp. on Permafrost Geophysics (5), Nov. 1978. N.R.C., Assoc. Committee on Geotechnical Res., Tech. Mem. 128, pp. 130-138.

Heinke, G.W. (1974) Arctic Waste Disposal. Task Force on Northern Oil Development. Rept. 74-10. 255 pp.

Heinke, G.W. and D. Prasad (1976) Disposal of Human Wastes in Northern Areas, in "Some Problems of Solid and Liquid Waste Disposal in the Northern Environment." Environmental Protection Service Rept. EPS 4-NW-76-2. Edmonton, Alberta, pp. 87-140.

Heinke, G.W. and D. Prasad (1977) Disposal of Human Wastes in Northern Areas. Dept. of Civil Eng., Univ. Toronto.

Hemming, J.E. (1984) Environmental Protection of Permafrost Terrain. Introduction and Case Study. Permafrost. 4th Int. Conf. Fairbanks. Final Proc. National Academy Press, Washington, pp. 113-115.

Henderson, J.D. (1980) Permafrost Mapping Along Transportation Corridors. Proc. Symp. on Permafrost Geophysics, Nov. 1978. N.R.C., Assoc. Committee on Geotechnical Res., Tech. Mem 128, pp. 130-138.

Hennion, F.B., and E.F. Lobacz (1973) Corps of Engineers Technology Related to Design of Pavements in Areas of Permafrost. Proc. 2nd Int. Conf. Permafrost, Yakutsk, USSR. North American Contribution. Nat. Acad. Sciences, Washington, pp. 658-664.

Hessler, V.P. and A.R. Franzke (1958) Earth-Potential Electrodes in Permafrost and Tundra. Arctic, 11, 211-217.

Hnatiuk, J. and E.E. Felzien (1985) Molikpaq - an Integrated Mobile Arctic Drilling Caisson. Proc. 17th Ann. Offshore Tech. Conf., Houston, Texas.

Hnatiuk, J. and B.D. Wright (1984) Ice Management to Support the Kulluk Drilling Vessel. J. Can. Pet. Tech. Sept.-Oct. 1984, 40-46.

Hoekstra, P., E. Chamberlain and T. Frate (1965) Frost Heaving Processes. U.S. Army Cold Regions Res. and Eng. Lab. Res. Rept. 176. 16 pp.

Hoekstra, P. and R.D. Miller (1965) Movement of Water in a Film Between Glass and Ice. U.S. Cold Regions Res. Eng. Lab. Res. Rept. 153. 8 pp.

Högbom, B. (1910) Einige illustrationen zu den geologischen Wirkungen des Frostes auf Spitzbergen. Uppsala Univ., Geol. Inst. Bull. 9, 41-59.

Högbom, B. (1914) Über die geologische Bedeutung des Frostes. Uppsala Univ., Bull. Geol. Inst. 12, 257-389.

Holdsworth, G. (1971) Calving from Ward Hunt Ice Shelf, 1961-1962. Can. J. Earth Sciences, 8, 299-305.

Hollingshead, G.W., L. Skjolingstad and L.A. Rundquist (1978) Permafrost Beneath Channels in the Mackenzie Delta, N.W.T., Canada. Proc. 3rd Int. Conf. Permafrost, Edmonton, 1, 406-412.

Hopkins, D.M. and P.V. Sellmann (1984) Subsea Permafrost. Introduction. Permafrost. 4th Int. Conf., Fairbanks. Final Proc. National Academy Press, Washington, pp. 73-74.

Howe, C.W. and F.P. Linaweaver (1967) The Impact of Price on Residential Water Demand and its Relation to System Design and Price Structure. Water Resources Res., 3, 13-32.

References

Hrudey, S.E. and S. Raniga (1981) Greywater Characteristics, Health Concerns and Treatment Technology, in D.W. Smith and S.E. Hrudey (eds.), Design of Water and Wastewater Services for Cold Climate Communities. Pergamon Press, Oxford, pp. 137-154.

Hsieh, Tzu-Ch'u, et al. (1975) Basic Features of the Glaciers of the Mt.Jolmo Lungma Region, Southern Part of the Tibet Autonomous Region, China. Scientia Sinica, 38, 106-130.

Huculak, N.A., J.W. Twach, R.S. Thomson and R.D. Cook (1979) Development of the Dempster Highway North of the Arctic Circle. Proc. 3rd Int. Conf. Permafrost, Edmonton. N.R.C., Ottawa. 947 pp. pp. 798-805.

Hudec, P.P. (1974) Weathering of Rocks in Arctic and Sub-Arctic Environment, in J.D. Aitken and D.J. Glass (eds.), "Canadian Arctic Geology", Geol. Assoc. of Canada, Can. Soc. Petroleum Geologists, Proc. Symp. on the Geology of the Canadian Arctic, pp. 313-335.

Hughes, O.L. (1972) Surficial Geology and Land Classification, Mackenzie Valley Transportation Corridor. Proc. Can. Northern Pipeline Res. Conf., Canada. N.R.C., Assoc. Committee Geotechnical Res., Tech. Mem. 104, pp. 17-24.

Hunt, L.A.C.O. (1978) Farming in the Territories. North-Nord, Jan. - Feb. 1978, 20-23.

Hunter, J.A.M. (1984) Geophysical Techniques for Subsea Permafrost Investigations. Permafrost. 4th Int. Conf., Fairbanks. Final Proc. National Academy Press, Washington, pp. 88-89.

Hunter, J., A. Judge, H. Macaulay, R. Good, R. Gagne and R.A. Burns (1976) The Occurrence of Permafrost and Frozen Subsea Materials in the Southern Beaufort Sea. Beaufort Sea Tech. Rept. 22. Dept. of the Environment, Victoria, B.C. 174 pp.

Huschke, R.E. (1959) Glossary of Meteorology. Amer. Met. Soc., Boston. 638 pp.

Hussey, K.M. (1962) Ground Patterns as Keys to Photointerpretation of Arctic Terrain. Proc. Iowa Acad. Science, 69, 332-341.

Isham, J. (1949) James Isham's Observations on Hudson's Bay,1743, and Notes and Observations on a Book Entitled a Voyage to Hudson's Bay in the Dobbs Galley, 1749. Ed. by E.E. Rich and A.M. Johnson. Champlain Soc. for Hudson's Bay Record Soc.

Ives, J.D. (1962) Iron Mining in Permafrost: Central Lanbrador-Ungava. Geog. Bull. 17, 66-77.

Ives, J.D. (1979) A Proposed History of Permafrost Development in Labrador-Ungava. Géographie Physique et Quaternaire, 33, 233-244.

Jackman, A.H. (1974) Highway Cut Stabilization in Areas of Permafrost and Ground Ice. Proc. Amer. Assoc. Geographers, 6, 29-32.

Jahn, A. (1976) Geomorphological Modelling and Nature Protection in Arctic and Subarctic Environments. Geoforum, 7, 121-137.

Jahns, H.O. (1984a) Subsea Permafrost and Petroleum Development. Permafrost. 4th Int. Conf., Fairbanks. Final. Proc. National Academy Press, Washington, pp. 90-92.

Jahns, H.O. (1984b) Pipeline Thermal Considerations. Permafrost. 4th Int. Conf., Fairbanks. Final Proc. National Academy Press, Washington, pp. 101-105.

Jahns, H.O., T.W. Miller, L.D. Power, W.P. Rickey, T.P. Taylor and J.A. Wheeler (1973) Permafrost Protection for Pipelines. Proc. 2nd Int. Permafrost Conf., Yakutsk, USSR. North American Contribution, Nat. Acad. Sciences, Washington, pp. 684-687.

Jeffrey, W.W. (1967) Forest Types Along Lower Liard River, Northwest Territories. Dept. of Forestry, Canada. Publication 1035. 103 pp.

Jeffries, M.O. (1983) Arctic Ice Shelf Studies, Spring 1983. Defence Res. Establishment Pacific, F.M.O., Victoria. Contractors' Report Series 83-27. 31 pp.

Jeffries, M. (1984) Mammoth Pads Circle the Arctic Basin. Geos, 13, 2-5. Energy, Mines & Resources, Ottawa.

References

Jenness, J.L. (1949) Permafrost in Canada, Origin and Distribution of Permanently Frozen Ground, with Special Reference to Canada. Arctic, 2, 1-27.

Jessup, R.W. (1960) The Stony Tableland Soils of the Southeastern Portion of the Australian Arid Zone and Their Evolutionary History. J. Soil Science, 11, 188-196.

Jian, S. and Z. Xianggong (eds.) (1982) Bibliography of Literature on China's Glaciers and Permafrost. Part 1: 1938-1979. Special Rept. 80-20. U.S. Army Corps of Engineers, Cold Regions Res. and Eng. Lab., Hanover, New Hampshire. 44 pp.

Johansson, S. (1914) Die Festigkeit der Bodenarten bei verschidenem Wassergehalt nebst Verschlag zu einer Klassifikation. Sveriges Geol. Undersökning Årsbok, 7, (arh. och uppsatser, ser. C, 256). 110 pp.

Johnson, D.L. and K.L. Hanson (1974) The Effects of Frost-heaving on Objects in Soil. Plains Anthropologist, 19, 81-98.

Johnson, D.L., D.R. Muhs and M.L. Barnhardt (1977) The Effects of Forst Heaving on Objects in Soils. II. Laboratory Experiments. Plains Anthropologist, 22(76), 133-147.

Johnson, E.G. and D.C. Esch (1977) Investigation and Analysis of the Paxson Roadway Icing. Proc. 2nd Int. Symp. Cold Regions Eng. J. L. Burdick and P. Johnson (eds.). Univ. Alaska, pp. 100-126.

Johnson, E.R. (1981) Buried Oil Pipeline Design and Operation in the Arctic - Lessons Learned on the Alyeska Pipeline. 37th Petroleum & Mech. Conf., Dallas, Texas.

Johnson, E.R. (1984) Performance of the Trans-Alaska Oil Pipeline. Permafrost. 4th Int. Conf., Fairbanks. Final Proc. National Academy Press, Washington, pp. 109-111.

Johnson, L. and L. Viereck (1983) Recovery and Active Layer Changes Following a Tundra Fire in Northwestern Alaska. Permafrost. 4th Int. Conf. Proc. National Academy Press, Washington, D.C., pp. 543-547.

Johnson, L.A. (1981) Revegetation and Selected Terrain Disturbances Along the Trans-Alaska Pipeline, 1975-1978. U.S. Army Cold Regions Res. and Eng. Lab. Rept. 81-12. 115 pp.

Johnson, P.R. (1971) Empirical Heat Transfer Rates of small, Long and Balch Thermal Piles and Convection Loops. Univ. Alaska, Inst. Arctic Environmental Eng., Fairbanks. Rept. 7102. 60 pp.

Johnson, T.C., R.L. Berg, K.L. Carey and C.W. Kaplar (1974) Roadway Design in Seasonal Frost Areas. U.S. Army, Cold Regions Res. & Eng. Lab., Tech. Rept. TR 259. 104 pp.

Johnston, A.V. (1964) Some Economic and Engineering Aspects of the Construction of New Railway Lines in Northern Canada, with Particular Reference to the Great Slave Railway. Proc. Inst. Civil Eng., 29, 571-588. Also Discussion, 32, Sept. 1965, 135-147.

Johnston, G.H. (1965) Permafrost Studies at the Kelsey Hydro-Electric Generating Station: Research and Instrumentation. N.R.C., Building Res. Div., Ottawa. Tech. Paper 178. 57 pp.

Johnston, G.H. (1966) Pile Construction in Permafrost. Proc. Int. Conf. Permafrost, Lafayette, Indiana. U.S. Nat. Acad. Sciences Publication 1287, pp. 371-374.

Johnston, G.H. (1969) Dykes on Permafrost, Kelsey Generating Station, Manitoba. Can. Geotechnical J., 6, 139-157.

Johnston, G.H. (1980) Permafrost and the Eagle River Bridge, Yukon Territory, Canada. Proc. Permafrost Eng. Workshop. N.R.C., Ottawa. Div. of Building Res., Tech. Mem 130, pp. 12-28.

Johnston, G.H. (Ed.) (1981) Permafrost: Engineering Design and Construction, Associate Committee on Geotechnical Research, National Research Council of Canada, Ottawa, Ontario, John Wiley & Sons (Canada) Ltd., Toronto, Ontario, 540 pp.

Johnson, G.H. (1982) Design and Performance of the Inuvik, N.W.T. Airstrip. The Roger J.E. Brown Memorial Volume, Proc. 4th Can. Permafrost Conf., Calgary, Alberta, March 1981, H.M. French, Ed., Associate Committee on Geotechnical Research, National Research Council of Canada, Ottawa, Ontario, pp. 577-585.

References

Johnston, G.H. (1983) Performance of an Insulated Roadway on Permafrost, Inuvik, N.W.T, in Permafrost: 4th Int. Conf. Proc., National Academy Press, Washington. 1524 pp. pp. 548-553.

Johnston, G.H., R.J.E. Brown and D.N. Pickersgill (1963) Permafrost Investigations at Thompson, Manitoba - Terrain Studies. N.R.C., Div. of Building Res., Ottawa. NRC 7568.

Johnston, G.H. and B. Ladanyi (1-72) Field Tests of Grouted Rod Anchors in Permafrost. Can. Geotechnical J., 9, 176-194.

Johnston, G.H. and B. Ladanyi (1974) Field Tests of Deep Power-Installed Screw Anchors in Permafrost. Can. Geotechnical J., 11, 348-358.

Johnston, G.H. and J.G. Macpherson (1981) Dams and Reservoirs, in G.H. Johnston (ed.), Chapter 9, Permafrost. Engineering Design and Construction. J. Wiley & Sons, New York, pp. 393-413.

Johnston, J.C. (1948) Planning and Operation of Hydraulic Stripping Plants. Unpublished thesis, Univ. Alaska.

Joseph, A.H. and S.L. Webster (1971) Techniques for Rapid Road Construction Using Membrane-Enveloped Soil Layers. U.S. Army Engineer Waterways Experiment Station, Instruction Rept. 5-71-1.

Judge, A.S. (1973a) Deep Temperature Observations in the Canadian North. North American Contribution, Permafrost. 2nd Int. Conf., Yakutsk, USSR. Nat. Academy Sciences, Washington, pp. 35-40.

Judge, A.S. (1973b) The Thermal Regime of the Mackenzie Valley: Observations of the Natural State. Environmental-Social Committee, Northern Pipelines Task Force on Northern Oil Development Rept. No. 73038. Information Canada R72-11973. 177 pp.

Judge, A.S. (1974) Occurrence of Offshore Permafrost in Northern Canada. J.C. Reed and J.E. Sater (eds.), "The Coast and Shelf of the Beaufort Sea," Proc. Symp. on Beaufort Sea Coastal and Shelf Res. Arctic Inst. of N. Amer. San Francisco, pp. 427-437.

Judge, A. (1982) Natural Gas Hydrates in Canada. Proc. 4th Can. Permafrost Conf., Calgary. N.R.C., Ottawa, 591 pp. pp. 320-328.

Jumikis, A.R. (1978) Graphs for Disturbance-Temperature Distribution in Permafrost Under Heated Rectangular Structures. Proc. 3rd Int. Conf. Permafrost, Edmonton. N.R.C., Ottawa. 947 pp. pp. 589-596.

Kaplar, C.W. (1965) Stone Migration by Freezing of Soil. Science, 149, 1520-1521.

Kaplar, C.W. (1969) Phenomena and Mechanism of Frost Heaving, in Highway Research Board Ann. Meeting, 49th, Washington, 1970 preprint. 44 pp.

Kaplar, C.W. (1971) Experiments to Simplify Frost Susceptibility Testing of Soils. U.S. Army Cold Regions Res. and Eng. Lab. Tech. Rept. 223. 21 pp.

Kaplar, C.W. (1974) Freezing Test for Evaluating Relative Frost Susceptibility Testing of Soils. U.S. Army Cold Regions Res. and Eng. Lab. Tech. Rept. TR 250. 36 pp.

Kastelic, W.R. (1982) Placer Deposits of Nome. Alaska Geographic, 9(4), 134-139.

Kelley, J.J. and T.A. Gosink (1984) Carbon Dioxide in the Arctic Atmosphere: Air-Sea and Air-Land Interaction, in J.H. McBeath (ed.), The Potential Effects of Carbon Dioxide - Induced Climatic Changes in Alaska. School of Agriculture & Land Resource Management, Univ. Alaska. Misc. Publ. 83-1, pp. 40-48.

Kellogg, W.W. (1984) Possible Effects of a Global Warming on Arctic Sea Ice, Precipitation, and Carbon Balance, in J.H. McBeath (ed.), The Potential Effects of Carbon Dioxide-Induced Climatic Changes in Alaska. School of Agriculture & Land Resources Management, Univ. Alaska. Misc. Publ. 83-1, pp. 59-66.

References

Kenyon, W.A. (1975) Tokens of Possession: The Northern Voyages of Martin Frobisher. Royal Ont. Museum, Toronto.

Kerfoot, D.E. (1972) Tundra Disturbance Studies of the Western Canadian Arctic. Dept. of Indian & Northern Affairs. Arctic Land Use Research 71-72-11.

King, M.S. and O.M. Garg (1980) Interpretation of Seismic and Resistivity Measurements in Permafrost in Northern Quebec. Proc. Symp. Permafrost Geophysics (No. 5), 1978. N.R.C. of Canada, Assoc. Committee on Geotechnical Res., Tech. Mem. 128, pp. 50-59.

Kinosita, S. (1967) Heaving Force of Frozen Soil. Proc. Int. Conf. on Low Temp. Science, 1, 1345-1350.

Kinosita, S., et al. (1979) Core Samplings of the Uppermost Layer in a Tundra Area. S. Kinosita (ed.), Joint Studies on Physical and Biological Environments in the Permafrost, Alaska and North Canada. Hokkaido Univ. Inst. of Low Temp. Science, Hokkaido. 149 pp.

Kivisild, H.R. and S.H. Iyer (1978) Mathematical and Physical Modelling of Ice, in: Int. Assoc. for Hydraulic Res., Symp. on Ice Problems, Lulea, Sweden. Proc., Part I. Ice Forces on Structures. 525 pp. pp. 379-391.

Klebesadel, L.J. and S.H. Restad (1981) Agriculture and Wildlife, are They Compatible in Alaska? Agroborealis, Jan., 15-23.

Kobayashi, R., H.J. Withrow, G.B. Williams and D.L. Katz (1951) Gas Hydrate Formation with Brine and Ethanol Solutions. Proc. 13th Ann. Convention, Natural Gasolene Assoc. Amer., pp. 27-31.

Kolodeznyi, P.A. and S.A. Arshinov (1970) The Technology of Pumping Antihydrate Inhibitor into Wells of the Messoyakha Field. Nauch. Tekh. Sb. Serv. Gazovoc Delo, 11, 9-

Konishchev, V.N. (1973) Kriogennoye vyretrivaniye, in: Akedemiya Nauk SSSR, Sektsiya Nauk o Zemle, Sibirskoye Otdeleniye, II. Mezhdunarodnaya Konferentsiya po Merzlotovedeniyu, Doklady; soobshcheniya. 3. Yakutsk, Yakutskoye Knizhnoye Izdatel'stvo. 102 pp.

Konishchev, V.N. (1978) Frost Weathering. F.J. Sanger (ed.), USSR Contribution. Permafrost. 2nd Int. Conf., Yakutsk. Nat. Acad. Sciences, Washington, pp. 176-181.

Konishchev, V.N., M.A. Fanstova and V.V. Rogov (1973) Cryogenic Processes as Reflected in Ground Microstructure. Biuletyn Peryglacjalny, 22, 213-219.

Konishchev, V.N., V.V. Rogov and G.N. Shehurina (1975) Cryogenic Transformation of Clayey Sediment Rocks. Fondation Française d'Études Nordiques, 6th Congrès International, "Les problèmes posés par la gélifraction. Recherches fondamentales et appliquées. Rept. 104.

Korpijaakko, E. and N.W. Radforth (1966) Aerial Photographic Interpretation of Muskeg Conditions at the Southern Limit of Permafrost. Proc. 11th Muskeg Res. Conf., Assoc. Committee on Geotechnical Res., N.R.C. of Canada, Tech. Mem. 87, pp. 142-151.

Kovacs, A. and D.S. Sodhi (1979) Shore Ice Pile-up and Ride-up. Workshop on Problems of Seasonal Sea Ice Zone, Naval Post Graduate School, Monterey, California.

Kovaleva, N.A., L.F. Mikheeva and M.I. Demina (1978) Waste Water Use for Feed Crop Land. U.S. Army, Cold Regions Res. and Eng. Lab., Technical Translation TL-673. 3 pp.

Kowalkowski, A. (1978) The Catena of Permafrost Soils in the Bayen-Nuurin-Khotnor Basin, Khangai Mountains, Mongolia. Proc. 3rd Int. Conf. Permafrost, Edmonton. N.R.C., Ottawa, 1, pp. 413-418.

Krebs, C.J. (1961) Population Dynamics of the Mackenzie Delta Reindeer Herd, 1938-1958. Arctic, 14, 91-100.

Kreig, R.A. (1985) Suggested Legend Terminology for Permafrost Mapping. Proc. Workshop on Permafrost Geophysics, Golden, Colorado. U.S. Army, CRREL, Special Rept. 85-5, pp. 41-47.

Kritz, M.A. and A.E. Wechsler (1967) Surface Characteristics, Effects on Thermal Regime: Phase II. U.S. Army, Cold Regions Res. and Eng. Lab., Tech. Rept. 189. 40 pp.
Krumme, O. (1935) Frost und Schnee in ihrer Wirking auf den Boden im Hochtaunus. Rhein-Mainische Forschungen 13. 73 pp.
Kry, P.R. (1977) Ice Rubble Fields in the Vicinity of Artificial Islands. Proc. 4th Int. Conf. on Port & Ocean Eng. under Arctic Conditions. Memorial Univ., Newfoundland.
Kry, P.R. (1980) Ice Forces on Wide Structures. Can. Geotechnical J., 17, 97-113.
Kudryavtsev, V.A., K.A. Kondrat'yeva and N.N. Romanovskiy (1978) Zonal'nyye i regional'-nyye zakonomernosti formiravaniya kriolitozony SSSR. (Zonal and Regional Patterns of Formation of the Permafrost Region in the USSR.) Proc. 3rd Int. Conf. Permafrost, Edmonton, 1, pp. 419-426, N.R.C., Ottawa.
Kudryavtsev, V.A., K.A. Kondvat'eva and N.N. Romanovskii (1980) Zonal and Regional Patterns of Formation of the Permafrost Region in the U.S.S.R. 3rd Int. Conf. Permafrost. Part 1. English translations of 26 of the Soviet papers. N.R.C. of Canada NRCC 18119. Ottawa, pp. 371-389.
Kurfurst, P.J. and D.F. Van Dine (1973) Terrain Sensitivity and Mapping, Mackenzie Valley Transportation Corridor. Geol. Sur. Can., Paper 73-1, Part B, pp. 155-159.
Lachenbruch, A.H. (1957) Thermal Effects of the Ocean on Perma-Frost. Bull. Geol. Soc. Amer., 68, 1515-1530.
Lachenbruch, A.H. (1962) Mechanics of Thermal Contraction Cracks and Ice Wedge Polygons in Permafrost. Geol. Soc. Amer., Special Paper 79. 69 pp.
Lachenbruch, A.H. (1968) Permafrost, in R.W. Fairbridge (ed.), Encyclopedia of Geomorphology. Reinhold Book Co., New York, pp. 833-838.
Lachenbruch, A.H. (1970) Thermal Considerations in Permafrost, in W.L. Adkison and M.M. Borsge (eds.) Geological Seminar on the North Slope of Alaska Proc. Los Angeles. Amer. Assoc. Petroleum Geologists, Pacific Section, Al-Rio. J1-2 and discussion J2-5.
Lachenbruch, A.H. and B.V. Marshall (1969) Heat Flow in the Arctic. Arctic, 22, 300-311.
Ladanyi, B. (1972) An Engineering Theory of Creep of Frozen Soils. Can. Geotechnical J., 9, 63-80.
Ladanyi, B. (1984) Design and Construction of Deep Foundations in Permafrost: North American Practice. Permafrost. 4th Int. Conf., Fairbanks. Final Proc. National Academy Press, Washington, pp. 43-50.
Lagerac, D. (1982) Cryogenetic Mounds as Indicators of Permafrost Conditions, Northern Quebec. Proc. 4th Can. Permafrost Conf., Calgary. N.R.C., Ottawa, pp. 43-48.
Lagov, P.A. and O. Yu. Parmuzina (1978) Ice Formation in the Seasonally Thawing Layer (field observations at the Ust'-Yenisey Research Station). (In Russian). In: General Geocryology. P.I. Melnikov. Nauka, Novosibirsk, U.S.S.R., pp. 56-59.
Lambe, T.W. and C.W. Kaplar (1971) Additives for Modifying the Frost Susceptibility of Soils. U.S. Army, Cold Regions Res. and Eng. Lab., Hanover, New Hampshire. Tech. Rept. 123, Pt. 1, 41 pp; Pt. 2, 41 pp.
Lambert, J.D.H. (1972) Botanical Changes Resulting from Seismic and Drilling Operations, Mackenzie Delta Area. Dept. of Indian and Northern Affairs, Ottawa. ALUR 71-72-12.
Langohr, R. and B. Van Vliet (1981) Properties and Distribution of Vistulian Permafrost Traces in Today Surface Soils in Belgium, with Special Reference to the Data Provided by the Soil Survey. Biuletyn Peryglacjalny, 28, 137-148.
Langway, C.C. Jr. (1967) Stratigraphic Analysis of a Deep Ice Core from Greenland. U.S. Army, CRREL. Res. Dept. 77. 132 pp.

References

Leffingwell, E. de K. (1915) Ground-ice Wedges; the Dominant Form of Ground-ice on the North Coast of Alaska. J. of Geol., 23, 635-654.

Lefroy, G.H., Sir (1889) Report upon the Depth of Permanently Frozen Soil in the Polar Regions, its Geographical Limits, and Relation to the Present Poles of Greatest Cold. Proc. Royal Geographical Soc., 8, 740-746.

Legget, R.F. (1941) Construction North of 54°. Eng. J., 24, 346-348.

Legget, R.F. (1965) Permafrost in North America, in Proc. Perma-Frost Int. Conf., 1963, Nat. Acad. Science, Washington, and N.R.C., pp. 2-7.

Legget, R.F., R.J.E. Brown and G.H. Johnston (1966) Alluvial Fan Formation Near Aklavik, Northwest Territories, Canada. Bull. Geol. Soc. Amer., 77, 15-29.

Leshchikov, F.N. and T.G. Ryashchenko (1973) Izmeneniye sostava i svoystv glinistykh gruntov pri promerzanii, in: Akademiya Nauk SSSR. Sektsiya Nauk o Zemle, Sibirskoye Otdeleniye. II. Mezhdunarodnaya Konferentsiya po Merzlotovedeniyu, Doklady i soobshcheniya, 3. Yakutsk, Yakutskoye Knizhnoye Izdatel'stvo. 102 pp.

Keshchikov, F.N. and T.G. Ryashchenko (1978) Changes in the Composition and Properties of Clay Soils During Freezing, in F.J. Sanger (ed.), USSR Contribution Permafrost 2nd Int. Conf., Yakutsk, USSR. Nat. Acad. Sciences, Washington, pp. 201-213.

Li, Yusheng, Wang Zhugui, Dai Jingbo, Cui Chenghan, He Changgen and Zhao Yunlong (1983) Permafrost Study and Railroad Construction in Permafrost Areas of China. Permafrost. 4th Int. Conf. Proc., Fairbanks. National Academy Press, Washington, pp. 707-714.

Liestøl, O. (1977) Pingos, Springs, and Permafrost in Spitsbergen. Norsk. Polarinstitutt Årbok, 1975, 7-29.

Lindsay, J.D. and W. Odynsky (1965) Permafrost in Organic Soils of Northern Alberta. Can. J. Soil Science, 45, 265-269.

Linell, K.A. (1973) Risk of Uncontrolled Flow from Wells Through Permafrost. Proc. 2nd Int. Conf. Permafrost, Yakutsk, USSR. North American Contribution. U.S. Nat. Acad. Sciences, Washington, pp. 462-468.

Linell, K.A. and G.H. Johnston (1973) Principles of Engineering Design and Construction in Permafrost Regions, in Permafrost: North American Contribution to the 2nd Int. Conf. on Permafrost, Yakutsk, USSR. Nat. Acad. Sciences, Washington, pp. 553-575.

Linell, K.A. and C.W. Kaplar (1959) The Factor of Soil and Material Type in Frost Action. U.S. Highway Res. Bd. Bull. 225, 81-126.

Linell, K.A. and C.W. Kaplar (1966) Description and Classification of Frozen Soils. Proc. Int. Conf. Permafrost, Lafayette, 1963. U.S. Acad. Sciences Publication 1287, pp. 481-487.

Linell, K.A. and E.F. Lobacz (1978) Some Experiences with Tunnel Entrances in Permafrost. Proc. 3rd Int. Conf. Permafrost, Edmonton. N.R.C., Ottawa, 1. 947 pp. pp. 813-819.

Livingstone, C.W. (1956) Excavations in Frozen Ground, Part 1, Explosion Tests in Keweenaw Silt. U.S. Army, Cold Regions Res. & Eng. Lab. (SIPRE). Hanover, New Hampshire. Tech. Rept. TR 30. 97 pp.

Livingston, H. and E. Johnson (1978) Insulated Roadway Subdrains in the Subarctic for the Prevention of Spring Icings. Proc. Conf. on Applied Techniques for Cold Environments, Anchorage, Alaska. Amer. Soc. Civil Engineers, 1, 513-521.

Lobacz, E.F. and K.S. Eff (1978) Storm Drainage Design Considerations in Cold Regions. Proc. Conf. on Applied Techniques for Cold Environments, Anchorage. Amer. Soc. Civil Engineers, 1, 474-487

Lobacz, E.F. and K.F. Eff (1981) Surface Drainage Design for Airfields and Heliports in Arctic and Subarctic Regions. U.S. Army, Cold Regions Res. and Eng. Lab., Special Rept. 81-2. Hanover, New Hampshire. 55 pp.

References

Lobacz, E.F. and W.F. Quinn (1966) Thermal Regime Beneath Buildings Constructed on Permafrost. Proc. 1st Int. Conf. Permafrost, Lafayette. Nat. Acad. Sciences, Washington. Publication 1287, pp. 247-252.
Long, E.L. (1966) The Long Thermopile. Proc. 1st Int. Conf. Permafrost, Lafayette. Nat. Acad. Sciences, Washington. Publication 1287, pp. 487-491.
Lopez, R.S., S.C.H. Cheung and R.A. Dixon (1984) The Canadian Program for Sealing Underground Nuclear Fuel Waste Vaults. Can. Geotechnical J., 21, 593-596.
Lötschert, W. (1972) Uber die vegetation frostgeformter Böden auf Island. Natur. u. Muscum 102, 1-12.
Luckman, B.H. (1972) Some Observations on the Erosion of Talus Slopes by Snow Avalanches in Surprise Valley, Jasper National Park, Alberta, in H.O. Slaymaker and H.J. McPherson (eds.), Mountain Geomorphology. Tantalus Research Ltd., Vancouver, pp. 85-100.
Luscher, U., W.T. Black and K. Nair (1975) Geotechnical Aspects of Trans-Alaska Pipeline. Amer. Soc. Civil Engineers, J. of the Transportation Eng. Div., 101(TE4, 669-680.
MacDonald, D.H., R.A. Pillman and H.R. Hopper (1960) Kelsey Generating Station Dam and Dykes. Eng. J., 43(10), 87-98.
MacFarlane, I.C (ed.) (1969) Muskeg Engineering Handbook. Univ. Toronto Press, Toronto. 320 pp.
Mackay, J.R. (1966) Segregated Epigenetic Ice and Slumps in Permafrost, Mackenzie Delta Area, N.W.T. Geog. Bull., 8, 59-80.
Mackay, J.R. (1970) Disturbances to the Tundra and Forest Tundra Environment in the Western Arctic. Can. Geotechnical J., 7, 420-432.
Mackay, J.R.(1971a) Ground Ice in the Active Layer and the Top Portion of the Permafrost. Proc. Seminar on the Permafrost Active Layer. N.R.C., Ottawa. Tech. Mem. 103, pp. 26-30.
Mackay, J.R. (1971b) Origin of Massive Icy Beds in Permafrost, Western Arctic Coast, Canada. Can. J. Earth Sciences, 8, 397-422.
Mackay, J.R.(1972a) The World of Underground Ice. Annals Assoc. Amer. Geog., 62, 1-22.
Mackay, J.R. (1972b) Offshore Permafrost and Ground Ice, Southern Beaufort Sea, Canada. Can. J. Earth Sciences, 9, 1550-1561.
Mackay, J.R. (1973a) Problems in the Origin of Massive Icy Beds, Western Arctic, Canada. North American Contribution, Permafrost. 2nd Int. Conf., Yakutsk, USSR. Nat. Acad. Sciences, Washington. 783 pp. pp. 223-228.
Mackay, J.R. (1973b) A Frost Tube for the Determination of Freezing in the Active Layer Above Permafrost. Can. Geotechnical J., 10, 392-396.
Mackay, J.R.(1974) Ice-Wedge Crocks, Garry Island, North-West Territories. Can. J. Earth Sciences, 11, 1366-1383.
Mackay, J.R. (1975) The Stability of Permafrost and Recent Climatic Change in the Mackenzie Valley, N.W.T. Geol. Sur. Can. Rept. of Activities. Pt. B. Paper 75-1B, pp. 173-176.
Mackay, J.R. (1976) Ice Wedges as Indicators of Recent Climatic Change, Western Arctic Coast. Geol. Sur. Can. Rept. of Activities. Part A. Paper 76-1A, pp. 233-234.
Mackay, J.R. (1977) Changes in the Active Layer from 1968-1976 as a Result of the Inuvik Fire. Geol. Sur. Can., Paper 77-1B, pp. 273-275.
Mackay, J.R. (1978) The Use of Snow Fences to Reduce Ice-Wedge Cracking, Garry Island, Northwest Territories. Geol. Sur. Can., Paper 78-1A, pp. 523-524.
Mackay, J.R. (1981) Active Layer Slope Movement in a Continuous Permafrost Environment, Garry Island, Northwest Territories, Canada. Can. J. Earth Sciences, 18, 1666-1680.
Mackay, J.R. (1983) Downward Water Movement into Frozen Ground, Western Arctic Coast, Canada. Can. J. Earth Sciences, 20, 120-134.

References

Mackay, J.R. (1984) The Frost Heave of Stones in the Active Layer Above Permafrost with Downward and Upward Freezing. Arctic & Alpine Permafrost, 16, 413-417.

Mackay, J.R. and R.F. Black (1973) Origin, Composition and Structure of Perennially Frozen Ground and Ground Ice: A Review. Permafrost: The North American Contribution to the 2nd Int. Conf., Nat. Acad. Sciences, Washington, pp. 185-192.

Mackay, J.R. and C. Burrows (1979) Uplift of Objects by an Upfreezing Ice Surface. Can. Geotechnical J., 17, 609-613.

Mackay, J.R. and Mackay, D.K. (1974) Snow Cover and Ground Temperatures, Garry Island, N.W.T. Arctic, 27, 288-296.

Mackay, J.R. and W.H. Mathews (1974a) Needle Ice Striped Ground. Arctic & Alpine Res., 6, 79-84.

Mackay, J.R. and W.H. Mathews (1974b) Movement of Sorted Stripes, the Cinder Cone, Garibaldi Park, B.C., Canada. Arctic & Alpine Res., 6, 347-359.

Mackay, J.R., J. Ostrick, C.P. Lewis and D.K. Mackay (1979) Frost Heave at Ground Temperatures Below 0°C, Inuvik, Northwest Territories. In "Current Research, Part A." Geol. Sur. Can., Paper 79-1A, pp. 403-405.

Makogan, Yu. F. (1974) Hydrates of Natural Gases (in Russian). Nedra Press, Moscow. (Mineral Resources). 208 pp.

Makogan, Yu. F. (1982) Perspectives for the Development of Gas-Hydrate Deposits. The Roger J.E. Brown Memorial Volume, Proc. 4th Can. Permafrost Conf., Calgary, Alberta, March 1981, H.M. French, Ed., Associate Committee on Geotechnical Research, National Research Council of Canada, Ottawa, Ontario, pp. 299-304.

Manikian, V. (1983) Pile Driving and Load Tests in Permafrost for the Kuparuk Pipeline System, in Permafrost. 4th Int. Conf. Proc., Fairbanks. National Academy Press, Washington. 1524 pp. pp. 804-810.

Maxwell, S.V., D. McGonigal and C. Graham (1983) Artificial Islands and Steel Structures in the Beaufort Sea, in: I. Dyer and C. Chryssostomidis (eds.), Arctic Technology and Policy. Proc. 2nd Ann. MIT Sea Grant College Program Lecture and Seminar. Hemisphere Publishing Co., Washington, pp. 97-111.

McBeath, J.H. (ed.) (1984) The Potential Effects of Carbon Dioxide Induced Climatic Changes in Alaska. School of Agriculture & Land Res. Management, Univ. Alaska. Misc. Publ. 83-1. 208 pp.

McCracken, C. and R.D. Revel (1982) Domestic and Commercial Vegetable Gardening in Dawson City, Yukon Territory, 1980. Arctic, 35, 395-402.

McGuire, R.L. (1982) Methane Hydrate Gas Production by Thermal Stimulation. Proc. 4th Can. Permafrost Conf., Calgary. N.R.C., Ottawa, pp. 356-362.

McHattie, R.L. and D.C. Esch (1983) Benefits of a Peat Underlay Used in Road Construction on Permafrost, in: Permafrost: 4th Int. Conf. Proc., National Academy Press, Washington. 1524 pp. pp. 826-831.

McRoberts, E.C., T.C. Law and T.K. Murray (1978) Creep Tests on Undisturbed Ice-Rich Silt. Proc. 3rd Int. Conf. Permafrost, Edmonton. N.R.C., Ottawa, 1, 539-545.

McRoberts, E.C. and N.R. Morgenstern (1974a) The Stability of Thawing Slopes. Can. Geotechnical J., 11, 447-469.

McRoberts, E.C. and N.R. Morgenstern (1974b) Stability of Slopes in Frozen Soil, Mackenzie Valley, N.W.T. Can. Geotechnical J., 11(4), 554-573.

McRoberts, E.C. and J.F. Nixon (1975) Reticulate Ice Veins in Permafrost, Northern Canada: Discussion. Can. Geotechnical J., 12, 159-162.

McTaggart Cowan, I. (1969) Ecology and Northern Development. Arctic, 22, 3-12.

References

McVee, C.V. and J.V. Tileston (1984) Regulatory Responsibilities in Permafrost Environments of Alaska from the Perspective of a Federal Land Manager. Permafrost. 4th Int. Conf., Fairbanks. Final Proc. National Academy Press, Washington, pp. 125-128.

Melnikov, P.I. and B.A. Olovin (1983) Permafrost Dynamics in the Area of the Viluy River Hydroelectric Scheme. In: Permafrost: 4th Int. Conf. Proc., National Academy Press, Washington. 1524 pp. pp. 838-842.

Metz, M.C. (1984) Pipeline Workpads in Alaska. Permafrost. 4th Int. Conf., Fairbanks. Final Proc. National Academy Press, Washington, pp. 106-108.

Metz, M.C., T.G. Krzewinski and E.S. Clarke (1982) The Trans-Alaska Pipeline System Workpad - An Evaluation of the Present Evidence. Proc. 4th Can. Permafrost Conf., Calgary. N.R.C., Ottawa. 591 pp. pp. 523-534.

Middleton, C. (1743) The Effects of Cold Together with Observations. . . at Price of Wales's Fort upon Churchill River in Hudson's Bay, North America. Philosophical Trans. Royal Soc., 42, 157-171. London.

Miller, J.M. (1971) Pile Foundations in Thermally Fragile Frozen Soils. Proc. Symp. on Cold Regions Eng. (1970); Amer. Soc. Civil Engineers, Univ. Alaska, 1, 34-72.

Miller, R.D. (1984) Thermal Induced Vegetation: A Qualitative Discussion. Permafrost. 4th Int. Conf., Fairbanks. Final Proc. National Academy Press, Washington, pp. 61-63.

Miyamoto, H.K. and G.W. Heinke (1979) Performance Evaluation of an Arctic Sewage Lagoon. Can. J. Civil Engineers, 6, 324-328.

Mollard, J.D. (1961) Guides for the Interpretation of Muskeg and Permafrost Conditions from Aerial Photographs. Proc. 6th Muskeg Res. Conf., Assoc. Committee on Soil and Snow Mechanics, N.R.C., Canada, Tech. Mem. 67, pp. 67-87.

Mollard, J.D. and J.A. Pihlainen (1966) Airphoto Interpretation Applied to Road Selection in the Arctic. Proc. 1st Permafrost Int. Conf., Lafayette, Indiana, Nat. Acad. Sciences, Washington, Publ. 1287, pp. 381-387.

Morack, J.L., H.A. MacAulay and J.A. Hunter (1983) Geophysical Measurements of subbottom Permafrost in the Canadian Beaufort Sea. Permafrost: Proc. 4th Int. Conf., Fairbanks. National Academy Press, Washington. 1524 pp. pp. 866-871.

Morgenstern, N.R. (1985) Recent Observations on the Deformation of Ice and Ice-Rich Permafrost, in M. Church & O. Slaymaker (eds.) Field and Theory. Lectures in Geocryology. Univ. British Columbia Press, pp. 133-153.

Morgenstern, N.R., S. Thomson and D. Mageau (1978) Explosive Cratering in Permafrost: State of the Art. Dept. of National Devence, Defence Res. Establishment, Suffield, Ralston, Alberta. Contract 8 SU 77-00015. 420 pp.

Müller, F. (1963) Englacial Temperature Measurements on Axel Heiberg Island, Canadian Arctic Archipelago. I.A.S.H. Publication #61. Commission on Snow and Ice, pp. 168-180.

Muller, S. (1943) Permafrost or Permanently Frozen Ground and Related Engineering Problems. Strategic Studies, 62. United States Army.

Muller, S.W. (1946) Permafrost. J.W. Edwards, Inc., Ann Arbor, Michigan. 231 pp.

Murfitt, A.W., W.B. McMullen, M. Baker and J.F. McPhail (1976) Design and Construction of Roads on Muskeg in Arctic and Subarctic Regions. Proc. 16th Muskeg Res. Conf. N.R.C., Ottawa. Assoc. Committee on Geotechnical Res., Tech. Mem. 116, pp. 152-185.

Murphy, R.S. and K.R. Ranganthan (1974) Bioprocesses of the oxidation ditch in a subarctic climate. In: Symposium on Wastewater Treatment in Cold Climates. Environmental Protection Service Report #EPS 3-WP-74-3. Ottawa, Ont., pp. 332-357.

References

Muschell, F.E. (1970) Pile Tips and Barbs Prevent Ice Uplift. Civil Eng., 40, 41-43.

Naldrett, D.N. (1982) Aspects of the Surficial Geology and Permafrost Conditions, Klondike Goldfields and Dawson City, Yukon Territory. Unpublished M.Sc. thesis, Dept. of Geography, Univ. Ottawa. 150 pp.

Nansen, F. (1922) Spitzbergen. F.A. Brockhaus, Leipzig, 3rd Ed. 327 pp.

National Geographic Society (1982a) The People's Republic of China. Map specially prepared to accompany the book, Journey into China.

National Geographic Society (1982b) Journey into China. Nat. Geog. Soc., Washington. 514 pp.

Naylor, L.L., R.O. Stern, W.C. Thomas and E.L. Arobio (1980) Socioeconomic Evaluation of Reindeer Herding in Northwestern Alaska. Arctic, 33, 248-272.

Neave, K.G. and P.V. Sellmann (1983) Seismic Velocities and Subsea Permafrost in the Beaufort Sea, Alaska. Permafrost: Proc. 4th Int. Conf., Fairbanks. National Academy Press, Washington. 1524 pp. pp. 894-898.

Nees, A. (1951) Pile Foundations for Large Towers on Permafrost. Amer. Soc. Civil Eng. Proc., 77(103). 10 p.

Nei Fengming (1983) Preventing Frost Damage to Railway Tunnel Drainage Ditches in Cold Regions. Permafrost: 4th Int. Conf. Proc., Fairbanks. National Academy Press, Washington, pp. 899-902.

Newbury, R.W., K.G. Beaty and G.K. McCullough (1978) Initial Shoreline Erosion in a Permafrost Affected Reservoir, southern Indian Lake, Canada. Proc. 3rd Int. Conf. Permafrost, Edmonton. N.R.C., Ottawa, 1, 427-433.

Nicholson, F.H. (1978a) Permafrost Modifications by Changing the Natural Energy Budget. Proc. 3rd Int. Conf. Permafrost, Edmonton, 1, pp. 61-67.

Nicholson, F.H. (1978b) Permafrost Distribution and Characteristics Near Schefferville, Quebec: Recent Studies. Proc. 3rd Int. Conf. Permafrost, Edmonton, 1, 427-434.

Nicholson, F.H. (1979) Permafrost Spatial and Temporal Variations Near Schefferville, Nouveau-Québec. Géographie Physique et Quaternaire, 33, 265-278.

Nicholson, F.H. and H.B. Granberg (1973) Permafrost and Snow-Cover Relationships Near Schefferville, Quebec. Permafrost: North American Contribution, 2nd Int. Permafrost Conf., Yakutsk, USSR. Nat. Acad. Sciences, Washington. Publ. 2115, pp. 151-158.

Nikiforoff, C. (1928) The Perpetually Frozen Subsoil of Siberia. Soil Science, 26, pp. 61-81.

Nixon, J.F. (1978) Geothermal Aspects of Ventilated Pad Design. Proc. 3rd Int. Conf. Permafrost, Edmonton. N.R.C., Ottawa, 1, 840-846.

Nixon, J.F. (1983) Geothermal Design of Insulated Foundations for Thaw Prevention, in Permafrost: 4th Int. Conf. Proc., Fairbanks. National Academy Press, Washington. 1524 pp. pp. 924-927.

Nixon, J.F. and E.C. McRoberts (1976) A Design Approach for Pile Foundations in Permafrost. Can. Geotechnical J., 13, 40-57.

Northern Miner (1974) Federal Government Backing Baffin Island's First Mine. Northern Miner, 60(15), 1-2.

Nowosad, F.S. (1963) Growing Vegetables on Permafrost. North-Nord, July-Aug., 42-45.

Oberman, N.G. (1974) Regional'nyye osobennosti merzloy zony Timano-Ural'skoy oblasti. Vysshikh uchcbn. zavedenni, Geologiya i razvedka Izv., 11, 98-103.

Odum, W.B. (1983) Practical Application of Underslab Ventilation System: Prudhoe Bay Case Study. In: Permafrost. 4th Int. Conf. Proc., Fairbanks. National Academy Press, Washington. 1524 pp. pp. 940-944.

Palazzo, A.J. (1976) Effects of Wastewater Application on the Growth and Chemical Composition of Forages. U.S. Army, Cold Regions Res. and Eng. Lab. Rept. CR 76-39. 8 pp.
Parameswaran, V.R. (1978) Adfreeze Strength of Frozen Sand to Model Piles. Can. Geotechnical J., 15, 494-500.
Parameswaran, V.R. and J.R. Mackay (1983) Field Measurements of Electrical Freezing Potentials in Permafrost Areas. Permafrost: 4th Int. Conf. Proc., Fairbanks. National Academy Press, Washington. 1524 pp. pp. 962-967.
Parmuzina, O.Yu. (1977) The Dynamics of the Cryogenic Structure of the Soils of the Active Layer. (in Russian). Herald of Moscow Univ., Geography Series 6, 91-94.
Parmuzina, O.Yu. (1978) Cryogenic Texture and Some Characteristics of Ice Formation in the Active Layer. (in Russian). In: Problems in Cryolithology, 7. A.I. Popov (ed.). Moscow Univ. Press, Moscow, pp. 141-164. Translated in: Polar Geography and Geology, July-Sept. 1980, 131-152.
Parmuzina, O.Yu. (1979) An Approach to the Question of the Redistribution of Moisture in Frozen Soil (according to full-scale observations). (in Russian). In: Problems of Cryolithology, 8, A.I. Popov (ed.). Moscow Univ. Press, Moscow, pp. 194-197.
Pavlov, A.V. and B.A. Olovin (1974) Artificial Thawing of Frozen Ground by Heat from Solar Radiation in Placer Working, Novosibirsk. (in Russian). Nauka Press, pp. 141-153.
Penner, E. (1960) The Importance of Freezing Rate in Frost Action in Soils. Proc. Amer. Soc. for Testing and Materials, 60, 1151-1165.
Penner, E. (1970) Frost Heaving Forces in Leda Clay. Can. Geotechnical J., 7, 8-16.
Penner, E. (1973) Frost Heaving Pressures in Particulate Materials. OECD. Symp. on Frost Action on Roads. Paris, 1, pp. 379-385.
Peretrukhin, N.A. and T.V. Potatueva (1983) Laws Governing Interactions Between Railroad, Roadbeds and Permafrost. In: Permafrost: 4th Int. Conf. Proc. National Academy Press, Washington. 1524 pp. pp. 984-987.
Perl'shtein, G.Z. and L.P. Savenko (1977) Approximate Calculation of Distribution of Heat Source Intensity in Electric Thawing of Frozen Ground. (in Russian). Proc. All-Union Res. Inst. of Gold & Rare Metals, 37, Magadan, 156-160.
Peters, C.J. and R. Perry (1981) The Formation and Control of Trihalomethanes in Water Treatment Processes. In: D.W. Smith & S.E. Hrudey (eds.), Design of Water and Wastewater Services for Cold Climate Communities, pp. 103-123.
Pettibone, H.C. (1973) Stability of an Underground Room in Frozen Gravel. Proc. 2nd Int. Conf. Permafrost, Yakutsk, USSR. North American Contribution. Nat. Acad. Sciences, Washington, pp. 699-706.
Péwé, T.L. (1954) Effect of Permafrost on Cultivated Fields, Fairbanks Area. U.S. Geol. Sur. Bull. 989-f, pp. 313-351.
Péwé, T.L. (1966a) Permafrost and its Effect on Life in the North. Oregon State Univ. Press, Corwallis, Oregon. 40 pp.
Péwé, T.L. (1966b) Ice-Wedges in Alaska - Classification, Distribution and Climatic Significance. Proc. 1st Int. Conf. Permafrost, Lafayette, Indiana. Nat. Acad. Sciences, Washington. Publ. 1287, pp. 76-81.
Péwé, T.L. (1970) Permafrost and Vegetation on Flood-Plains of Sub-Arctic Rivers (Alaska), a Summary. In: Ecology of the Subarctic Regions. UNESCO, Paris, pp. 141-142.
Philberth, K. and B. Federer (1971) On the Temperature Profile and the Age Profile in the Central Part of Cold Ice Sheets. J. of Glaciology, 10, 3-14.
Phukan, A. (1983) Long-Term Creep Deformation of Roadway Embankment on Ice-Rich Permafrost. In: Permafrost: 4th Int. Conf. Proc. National Academy Press, Washington. 1524 pp. pp. 994-999.

References

Pihlainen, J.A. (1951) Building Foundations on Permafrost, Mackenzie Valley, N.W.T. N.R.C., Div. Building Res., DBR 22.

Pihlainen, J.A. (1959) Pile Construction in Permafrost. Amer. Soc. Civil Eng., J. Soil Mechanics, Foundation Div., 85 (SM6), Pt. I, 75-95.

Pihlainen, J.A. (1962) Inuvik, N.W.T., Engineering Site Information. N.R.C., Div. of Building Res., Ottawa, NRC6757.

Pihlainen, J.A. and G.H. Johnston (1963) Guide to a Field Description of Permafrost. N.R.C.C. Assoc. Committee on Snow and Soil Mechanics, Tech. Mem. 79 (N.R.C. Publ. 7576). 23 pp.

Pikulevich, L.D. (1963) Freezing Processes and Changes in Moisture Content in Seasonally Thawing Soils (in Russian). Merzlotnye Issledovaniya, 3, 158-167.

Pilkington, C.R., B.D. Wright, L.G. Spedding and B.W. Danielewicz (1983) A Short Summary of the Physical Environment of the Beaufort Sea and its Effect on Offshore Operations, Gulf, Esso, Dome. Calgary.

Pissart, A. (1969) Le méchanism périglaciaire dressant les pierres dans le sol. Resultats d'expériences. Academie de Science, Paris, Comptes Rendus 268, pp. 3015-3017.

Pissart, A. (1973) Résultats d'expériences sur l'action du gel dans le sol. Biuletyn Peryglacjalny, 23, 101-113.

Pissart, A. (1969) Le méchanism périglaciaire dressant les pierres dans le sol. Résultats d'expériences. Academic de Science, Paris, Comptes rendus, 268, pp. 3015-3017.

Poley, D.F. (1982) A detailed Study of a Submerged Pingo-Like Feature in the Canadian Beaufort Sea, Arctic, Canada. Unpublished B.Sc. thesis, Dalhousie Univ., Halifax, Nova Scotia. 93 pp.

Popov, B.I., N.F. Savko, N.M. Tupitsya, Yu.K. Kleer and A.S. Plotskii (1980) Construction of Earth Embankments for Highways in Western Siberia. English Translations, Part II. Proc. 3rd Int. Conf. Permafrost, N.R.C., Ottawa, pp. 269-281.

Porkhaev, G.V., R.L. Valershtein, V.N. Eroshenko, A.L. Mindich, Yu. S. Mirenburg, V.D. Ponomurev and L.N. Khrustalev (1980) Construction by the Method of Stabilizing Perennially Frozen Foundation Soils. English Translations, Part II. Proc. 3rd Int. Conf. Permafrost. N.R.C., Ottawa, pp. 283-296.

Porsild, A.E. (1929) Reindeer Grazing in Northwest Canada. Dept. of the Interior, Northwest Territories and Yukon Branch, Ottawa. 46 pp.

Potts, A.S. (1970) Frost Action in Rocks: Some Experimental Data. Trans. Inst. Brit. Geographers, 49, 109-124.

Prasad, D. and G.W. Heinke (1981) Disposal and Treatment of Concentrated Human Wastes. In: D.W. Smith and S.E. Hrudey (eds.) Design of Water and Wastewater Services for Cold Climate Communities. Pergamon Press, Oxford, pp. 125-135.

Pressman, A.E. (1963) Comparison of Aerial Photographic Analysis with Investigations in Arctic Canada. Photogrammetric Eng., 29, 242-252.

Price, L.W. (1970) Up-heaved Blocks: A Curious Feature of Instability in the Tundra. Proc. Assoc. Amer. Geog., 2, 106-110.

Price, L.W. (1971) Vegetation, Microtopography, and Depth of Active Layer on Different Exposures in Subarctic Alpine Tundra. Ecology, 52, 638-647.

Protas'eva, I.V. (1959) Primeneniye Aerometodov Pri Geokriologicheskikh Issledovaniyakh. (Applications of Aeromethods to Geocryological Investigations). 7th Inter-departmental Permafrost Conf., Acad. Sciences, USSR, pp. 85-87.

References

Protas'eva, I.V. (1961) Primeneniya Aerofotos emki V. Izuchenii Mnogoletnemerzlykh Gornykh Porod (Application of Aerial Surveying in the Study of Permafrost). Trudy Igarskoy Nauchno-Issledovatel'skay Merzlotnoy Stantsiya, (Transactions of the Igarka Permafrost Research Station), 2, pp. 75-90.

Protas'eva, I.V. (1967) Aerometody v Geokriologii (Aerial Methods in Geocryology). Permafrost Inst., Siberian Div., Acad. Sciences, USSR. 195 pp.

Pryer, R.W.J. (1966) Mine Railroads in Labrador-Ungava. Proc. Int. Conf. Permafrost, Lafayette, Indiana. Nat. Acad. Sciences, Washington, Publ. 1287, pp. 503-508.

Pufahl, D.E. (1976) The Behavior of Thawing Slopes in Permafrost. Unpublished Ph.D. thesis, Dept. Civil Eng., Univ. Alberta. 345 pp.

Pufahl, D.E. and N.R. Morgenstern (1979) Stabilization of Planar Landslides in Permafrost. Can. Geotechnical J., 16, 734-747.

Pufahl, D.E., M.R. Morgenstern and W.D. Roggensack (1974) Observations on Recent Highway Cuts in Permafrost. Environmental-Social Program, Northern Pipelines, Task Force on Northern Oil Development, Rept. 74-32. 53 pp.

Quong, J.Y.C. (1971) Highway Construction and Permafrost with Special Reference to the Active Layer. Proc. Sem. on the Permafrost Active Layer, Vancouver, B.C. N.R.C., Ottawa, Assoc. Committee on Geotechnical Res., Tech. Mem. 103, pp. 50-53.

Radd, F.J. and D.H. Oertle (1973) Experimental Pressure Studies of Frost Heave Mechanisms and the Growth-Fusion Behaviour of Ice. Proc. 2nd Int. Conf. Permafrost, Yakutsk, USSR. North American Contribution. Nat. Acad. Sciences, Washington, pp. 377-384.

Radforth, J.R. (1972) Analysis of Disturbance Effects of Operation of Off-road Vehicles on Tundra. Dept. of Indian and Northern Affairs, Ottawa. ALUR 71-72-13.

Radforth, J.R. (1973) Immediate Effects of Wheeled Traffic on Tundra During the Summer. Dept. of Indian Affairs and Northern Development. Arctic Land Use Res. program, 72-73-12. 32 pp.

Radforth, N.W. (1963) The Ice Factor in Muskeg. Proc. 1st Can. Conf. Permafrost. N.R.C., Assoc. Committee on Soil and Snow Mechanics. Tech. Mem. 76, pp. 57-78.

Rampton, V.N. and J.R. Mackay (1971) Massive Ice and Icy Sediments Throughout the Tuktoyaktuk Peninsula, Richards Island, and Nearby Areas, District of Mackenzie. Geol. Sur. Can., Paper 71-21. 16 pp.

Ramsier, R.O., M.R. Vant and L.D. Arsenault (1975) Distribution of Ice Thickness in the Beaufort Sea. Beaufort Sea Project Tech. Rept. 30. Institute of Ocean Sciences, Sidney, B.C. 98 pp.

Ray, R.G. (1960) Aerial Photographs in Geologic Interpretation and Mapping. U.S. Geol. Sur., Prof. Paper 373. 230 pp.

Reed, R.E. (1966) Refrigeration of a Pipe Pile by Air Circulation. U.S. Army, Cold Regions Res. & Eng. Lab., Tech. Rept. 156. 19 pp.

Reid, G.D. (1974) The Impact of the Northern Environment on Highway Construction. Bull. Amer. Assoc. Cost Engineers, 16 (2), 43-47.

Reid, R.L., J.S. Tennant and K.W. Childs (1975) The Modelling of a Thermosyphon Type Permafrost Protection Device. Amer. Soc. Mechanical Eng., J. Heat Transfer, August, 382-386.

Research Notes (1982) Solar Assisted Culvert Thawing Device. State of Alaska Dept. of Transport & Public Facilities, Res. Notes 2(1). 2 pp.

References

Richardson, J. (1839) Notice of a Few Observations which it is Desirable to Make on the Frozen Soil of British North America. J. Royal Geog. Soc., 9, 117-120.

Richardson, J. (1851) A Journey of a Boat-Voyage Through Rupert's Land and the Arctic Sea, in Search of the Discovery Ships Under Command of Sir John Franklin. London. Longman, Brown, Green, and Longmans.

RIGCOR (1975) Permafrost. Res. Inst. of Glaciology, Cryopedology and Desert Res., Academia Sinica, Lanchou, China. Can. Inst. for Scientific and Technical Information. Tech. translation #2006, The Canada Institute for Scientific and Technical Information, National Research Council of Canada, Ottawa, 1981. 146 pp., pp. 1-44.

Roberts, B. (1942) The Reindeer Industry in Alaska. Polar Record, 24, 568-572.

Robin, G. de Q. (1972) Polar Ice Sheets: A Review. Polar Record, 100, 5-22.

Robson, J. (1752) Account of Six Years Residence in Hudson's Bay from 1733-36 and 1744-47. London. J. Payne and J. Bouquet.

Roggensack, W.D. (1977) Geotechnical Properties of Fine-Grained Permafrost Soils. Unpublished Ph.D. Thesis, Dept. of Civil Eng., Univ. Alberta, Edmonton. 423 pp.

Römkens, M.J.M. (1969) Migration of Mineral Particles in Ice with a Temperature Gradient. Unpublished Ph.D. Thesis, Cornell Univ. 109 pp.

Römkens, M.J.M. and R.D. Miller (1973) Migration of Mineral Particles in Ice with a Temperature Gradient. J. Colloid and Interface Science, 42, 103-111.

Rook, J.J. (1974) Formation of Haloforms During Chlorination of Natural Waters. J. Water Treat. Exam., 23, 234-243.

Roscow, J. (1977) 800 Miles to Valdez. Prentice-Hall Inc., New York. 277 pp.

Rosendahl, G.P. (1981) Alternative Strategies Used in Greenland. In: D.W. Smith & S.E. Hrudey (eds.), Design of Water and Wastewater Services for Cold Climate Communities. Pergamon Press, Oxford, pp. 3-16.

Rouse, W.R. and K.A. Kershaw (1971) The Effects of Burning on Heat and Water Regimes of Lichen-Dominated Subarctic Surfaces. J. Arctic & Alpine Res., 3, 291-304.

Rowley, R.K., G.H. Watson and B. Ladanyi (1973) Vertical and Lateral Pile Load Tests in Permafrost. Proc. 2nd Int. Conf. Permafrost, Yakutsk, USSR. North American Contribution. Nat. Acad. Sciences, Washington, pp. 712-721.

Ruhe, P.V. (1983) Aspects of Holocene Pedology in the United States. Chapter 2. In: H.E. Wright (ed.), The Holocene. Vol. 2, Late Quaternary Environments of the United States. Univ. Minnesota Press, pp. 12-25.

Sager, R.C. (1951) Aerial Analysis of Permanently Frozen Ground. Photogrammetric Eng., 17, 551-571.

Sanger, F.J. (1969) Foundations of Structures in Cold Regions. U.S. Army, Cold Regions Res. and Eng. Lab., Monograph M111-C4. 91 pp.

Sayles, F.H. (1984) Design and Performance of Water-Retaining Embankments in Permafrost. Permafrost. 4th Int. Conf., Fairbanks. Final Proc. National Academy Press, Washington, pp. 31-42.

Schaefer, D. (1973) MESL Construction at Fort Wainwright, Alaska. U.S. Army, Cold Regions Res. and Eng. Lab, Tech. Rept. (unpublished).

Schmid, J. (1955) Der Bodenfrost als morphologischer Faktor. Heidelburg. Dr. Alfred Hüthig Verlag. 144 pp.

Schunke, E. (1975) Die periglazialerscheinungen Islands in abhängigkeit von Klima und Substrat. Akad. Wiss. Göttingen Abh., Math.-Phys. Kl. Folge, 30(3), 273 pp.

Scott, R.F. (1969) The Freezing Process and Mechanics of Frozen Ground. United States Army Corps of Engineers, Cold Regions Res. and Eng. Lab. Cold Regions Science and Engineering Monograph II-D1. 69 pp.

Scott, W.J. and J.R. Mackay (1977) Reliability of Permafrost Thickness Determination by D.C. Resistivity Sounding. Proc. Symp. Permafrost Geophysics (1976), Vancouver, Canada. N.R.C., Assoc. Committee for Geotechnical Res., Tech. Mem. 119, pp. 25-38.

Sellman, P.V. and D.M. Hopkins (1984) Subsea Permafrost Distribution on the Alaskan Shelf. Permafrost. 4th Int. Conf., Fairbanks. Final Proc. National Academy Press, Washington, pp. 75-82.

Sellman, P.V., J.D. McNeill and W.J. Scott (1974) Airborne E-Phase Resistivity Surveys of Permafrost, Central Alaska and Mackenzie River Areas. Proc. Symp. Permafrost Geophysics, Calgary, Alberta. N.R.C.C., Assoc. Committee on Geotechnical Res., Tech. Mem 113, pp. 67-71.

Semmel, A. (1969) Verwitterungs- und Abtragungserscheinungen in rezentum Periglazialgebieten (Lappland und Spitzbergen). Würzburger Geographische Arbeiten, 26. 82 pp.

Seppälä, M. (1982) An Experimental Study of the Formation of Palsas. In: Proc. 4th Can. Permafrost Conf., Calgary. N.R.C., Ottawa, pp. 36-42.

Shabad, T. (1982) Soviet Announces Agricultural Program for 1980's Soviet Geography, 22, 463-465.

Shah, V.K. (1978) Protection of Permafrost and Ice Rich Shores, Tuktoyaktuk, N.W.T., Canada. Proc. 3rd Int. Conf. Permafrost, Edmonton. N.R.C., Ottawa, 1, 871-876.

Shearer, J.M., R.F. Macnab, B.R. Pelletier and T.B. Smith (1971) Submarine Pingos in the Beaufort Sea. Science, 176, 816-818.

Shi Yafenq, Li Jijun and Xie Zhichu (1979) The Uplift of the Qinghai-Xiang Plateau and its Effect on China During Ice Age. J. Glaciology & Cryopedology, 1, 6-11.

Shi Yafeng and Mi Daheng (1983) A Comprehensive Map of Snow, Ice and Frozen Ground in China (1:4,000,000). Permafrost. 4th Int. Conf. Proc. National Academy Press, Washington, pp. 1148-1151.

Shu Daode and Huang Xiaoming (1983) Design and Construction of Cutting at Sections of Thick-Layer Ground Ice. In: Permafrost: 4th Int. Conf. Proc. National Academy Press, Washington. 1524 pp. pp. 1152-1156.

Skaven-Haug, S. (1959) Protection Against Frost Heaving on the Norwegian Railways. Geotechnique, 9(3).

Slaughter, C.W. (1982) Occurrence and Recurrence of Aufeis in an Upland Taiga Catchment. Proc. 4th Can. Permafrost Conf., Calgary. N.R.C., Ottawa, pp. 182-188.

Slipchenko, W. (1972) Siberia, 1971. A Report on a Visit of the Honorable Jean Chrétien, Minister of Indian Affairs and Northern Development and Official Delegation to the Soviet Union. Information Canada. 124 pp.

Slupsky, J.W. (ed.) (1976) Some Problems of Solid and Liquid Waste Disposal in the Northern Environment. Environmental Protection Ser., Rept. EPS-4-NW-76-2. 230 pp. (Series of 4 reports).

Smith, D.W., and P.W. Given (1981) Treatment Alternatives for Dilute, Low Temperature Wastewater. In: D.W. Smith and S.E. Hrudey (eds.), Design of Water and Wastewater Services for Cold Climate Communities. Pergamon Press, Oxford, pp. 165-179.

Smith, D.W. and G.W. Heinke (1981) Cold Climate Environmental Engineering - an Overview. In: D.W. Smith and S.E. Hrudey (eds.), Design of Water and Wastewater Services for Cold Climate Communities, pp. 3-16.

Smith, D.W. and S.E. Hruday (eds.) (1981) Design of Water and Wastewater Services for Cold Climate Communities. Pergamon Press, Oxford. 184 pp.

References

Smith, D.W., S. Reed, J.J. Cameron, G.W. Heinke, F. James, B. Reid, W.L. Ryan and J. Scribner (1979) Cold Climate Utilities Delivery Design Manuel. Environmental Protection Service Rept. EPS 3-WP-79-2, Ottawa, Ont. 650 pp.

Smith, M.W. (1975) Microclimatic Influences on Ground Temperatures and Permafrost Distribution, Mackenzie Delta, Northwest Territories. Can. J. Earth Sciences, 12, 1421-1438.

Smith, M.W. (1985) Models of Soil Freezing. In: M. Church and O. Slaymaker (eds.), Field and Theory. Lectures in Geocryology. Univ. of British Columbia Press, pp. 96-120.

Smith, N. and R.L. Berg (1973) Encountering Massive Ground Ice During Road Construction in Central Alaska. Proc. 2nd Int. Conf. Permafrost, Yakutsk, USSR. North American Contribution. Nat. Acad. Sciences, Washington, pp. 736-745.

Smith, N. and D.A. Pazsint (1975) Field Test of a MESL (membrane-enveloped soil layer) Road Section in Central Alaska. U.S. Army Cold Regions Res. and Eng. Lab. Tech. Rept. 260. 40 pp.

Snodgrass, M.P. (1971) Waste Disposal and Treatment in Permafrost Areas: a Bibliography. U.S. Dept. of the Interior, Office of Library Services, Washington, Bibliography Series 22. 29 pp.

Sobczak, L. (1977) Ice Movements in the Beaufort Sea, 1973-1975, ERTS. J. Geophysical Res., 82, 1413-1418.

Solbraa, K. (1971) The Durability of Bark in Road Construction. Royal Norwegian Council on Science and Industry, Dec. 1971, 5.

Solov'yev, P.A. (1962) Alasnyy rel'yef Centralnoy Yakutii ago proiskhozhdeniye. In: Mugoletnemerzlyye porody i soputstvuyushehiye im Yavleniya na territorii Yakutskoy SSR. Moskva, Izdatel'stvo AN SSR.

Solov'yev, P.A. (1973a) Alas Thermokarst Relief of Central Yakutia. 2nd Int. Permafrost Conf., Yakutsk, USSR. Guidebook. Yakutsk. USSR Acad. Sciences, Section Earth Science, Siberian Div. 48 pp.

Solov'yev, P.A. (1973b) Thermokarst Phenomena and Landforms Due to Frost Heaving in Central Yakutia. Biuletyn Peryglacjalny, 23, 135-155.

Sørensen, T. (1935 Bodenformen und Pflanzendecke in Nordost-grönland. (in German). Meddelelser om Grønland, 93. 69 pp.

Spedding, L.G. (1974) The Extent and Growth Pattern of Landfast Ice in the Southern Beaufort Sea - Winter 1972/73. Esso Resources IPRT-2ME-74. Arctic Petroleum Operators Assoc. Project 54. 42 pp.

Spedding, L.G. (1975) Landfast Ice Movement in Mackenzie Delta 1973/74. Esso Resources IPRT-26ME-72. Arctic Petroleum Operators Assoc. Project 83. 325 pp.

Spedding, L.G. (1978) Statistics on Beaufort Sea Summer Ice Cover for Ice/Structure Collision Assessment. Pallister Resource Management Ltd. Esso Resources Internal Rept. IPRT-2ME-78. 113 pp.

Spedding, L.G. (1979) Landfast and Shear Zone Ice Conditions in the Southern Beaufort Sea - Winter 1977/78. Pallister Resource Management Ltd. Esso Resources IPRT-16ME-79. 215 pp.

Springer, M.E. (1958) Desert Pavement and Vesicular Layer of Some Soils in the Desert of the Lahontan Basin, Nevada. Proc. Soil Science Soc. Amer., 22, 63-66.

Stanek, Z. (1970) Acclimatization of Cultivated Plants on the Northern Limit of Agriculture in the USSR. Arctic, 23, 51-55.

Stanley Associates Engineering, Ltd. (1979) Dawson City, Yukon, Utilities Pre-design Report. Prepared for the Govt. of the Yukon Territory, Whitehorse, Yukon Territory.

Stanley, J.M. and J.E. Cronin (1983) Investigations and Implications of Subsurface Conditions Beneath the Trans Alaska Pipeline in Atigun Pass. In: Permafrost: 4th Int. Conf. Proc., National Academy Press, Washington. 1524 pp. pp. 1188-1193.

Stoeckler, E.G. (1948) Identification and Evaluation of Alaskan Vegetation from Airphotos with Reference to Soil, Moisture, and Permafrost Conditions. U.S. Army Corps of Engineers, St. Paul District. 102 pp.

Stoll, R.D. (1974) Effects of Gas Hydrates in Sediments. In: I.R. Kaplan (ed.), Natural Gases in Marine Sediments. Plenum Press, New York, pp. 235-248.

Straughn, R.O. (1972) The Sanitary Landfill in the Subarctic. Arctic, 25, 40-48.

Strilchuk, A.R. (1977) Ice Pressure Measurement, Netserk F-40, 1975-76. Esso Resources, Canada, Ltd. APOA Project 105.

Stromquist, L. (1973) Geomorfologiska studiar av blockfält i narra Skandinavien. Uppsala Univ., Naturgeografiska Institut, Avdelningen für Naturgeografi, UNGI Rapport 22. 161 pp.

Sumgin, M.I. (1927) Vechnaya merzlota pochv v pradelakh SSSR. Vladivostok. 372 pp.

Sweet, L. (1982) Solar Assisted Culvert Thawing Device. Res. Notes, State of Alaska, Dept. of Transportation and Public Facilities, Fairbanks, 2(1).

Swinzow, G.K. (1963) Tunneling in Permafrost. U.S. Army Cold Regions Res. and Eng. Lab., Hanover, N.H., Tech. Rept. 93, pp. 14-27.

Sykes, D.J. (1971) Effects of Fire and Fire Control on Soil and Water Relations in Northern Forests - A Preliminary Review. Proc. Fire in the Northern Environment - A Symposium. College, Fairbanks, Alaska, pp. 37-44.

Taber, S. (1918) Ice Forming in Clay Soils will Lift Surface Weights. Eng. News-Rec., 80, 262-263.

Taber, S. (1929) Frost Heaving. J. Geology, 37, 428-461.

Taber, S. (1930a) The Mechanics of Frost Heaving. J. Geology, 38, 303-317.

Taber, S. (1930b) Freezing and Thawing of Soils as Factors in the Destruction of Road Pavements. Public Roads, 11, 113-132.

Taber, S. (1953) Origin of Alaska Silts. Amer. J. Science, 251, 321-326.

Tart, R.G. Jr. (1983) Winter Constructed Gravel Islands. In: Permafrost: 4th Int. Conf. Proc. National Academy Press, Washington. 1524 pp. pp. 1233-1238.

Taylor, A. and A.S. Judge (1980) Permafrost Studies in Northern Quebec. Proc. Symp. on Permafrost Geophysics, Calgary. W.J. Scott & R.J.E. Brown (eds.). Assoc. Committee on Geotechnical Res., N.R.C., Ottawa. Tech. Mem. 128, pp. 94-102.

Tedrow, J.C.F. (1966) Polar Desert Soils. Soil Science Soc. Amer. Proc., 30, 381-387.

Terzaghi, K. (1952) Permafrost. Boston Soc. Civil Engineers J., 39, 1-50.

Thomas, H.P. and J.E. Ferrell (L983) Thermokarst Features Associated with Buried Sections of the Trans-Alaska Pipeline. Permafrost: Proc. 4th Int. Conf., Fairbanks. National Academy Press, Washington. 1524 pp. pp. 1245-1250.

Thomas, H.P., E.R. Johnson, J.M. Stanley, J.A. Shuster and S.W. Pearson (1982) Pipeline Stabilization Project at Atigun Pass. Proc. 3rd Int. Symp. Ground Freezing, Hanover, New Hampshire. U.S. Army Cold Regions Res. and Eng. Lab.

Thompson, E.G. and F.H. Sayles (1972) In situ Creep Analysis of a Room in Frozen Soil. Amer. Soc. Civil Eng., J. Soil Mechanics and Foundation Division, No. SM6, 899-915.

Thomson, S. (1963) Icings on the Alaska Highway. Proc. 1st Int. Conf. Permafrost, Lafayette, Indiana. Nat. Acad. Sciences, Washington. Publ. 1287, pp. 526-529.

Thoroddsen, Th. (1913) Polygonboden und 'Thufur' auf Island. Petermanns Mitt., 59, 253-255.

References

Thoroddson, Th. (1914) An Account of the Physical Geography of Iceland with Special Reference to the Plant Life. In: L. Kolderup-Rosenvinge and E. Warming (eds.), The Botany of Iceland, vol. 1, 1912-1918. Copenhagen, J. Frimodt; London, John Weldon. 675 pp. pp. 187-343.

Tobiasson, W. (1973) Performance of the Thule Hanger Soil Cooling Systems. Proc. 2nd Int. Conf. Permafrost, Yakutsk, USSR. North American Contribution. Nat. Acad. Sciences, Washington, pp. 752-758.

Tolstikhin, N.I. (1940) The Regime of Ground and Surface Waters in the Region of Permafrost Distribution. (in Russian). In: M.I. Shumgin et al., General Geocryology. Moscow. Akad. Nauk. SSSR. 340 pp.

Tricart, J. (1970) Convergence de phénomènes entre l'action du gel et celle du sel. Acta Geographica Lodziensia, 24, 425-436.

Trofimuk, A.A., Yu. F. Makogon and N.M. Chemakin (1980) Natural Gas Hydrates of North West Siberia. Geologiya i Geofizika; Izdatelstvo Nauka, 9, 3-9.

Troll, C. (1944) Strukturböden, Solifluktion und Frostklimate der Erde. Geologische Rundschau, 34, 545-694.

Troll, C. (1958) Structure Soils, Solifluction, and Frost Climates of the Earth. U.S. Army, Corps of Engineers, Snow, Ice and Permafrost Res. Establishment. Translation 43. 121 pp.

Trupak, N.G. (1970) Construction of Earth Dams on Permafrost Soils. Hydrotechnical Construction (U.S.S.R.), 9, 8-11. Translated by Consultants Bureau, New York, N.Y., for the Amer. Soc. Civil Engineers.

Tsytovich, N.A. (1957) The Fundamentals of Frozen Ground Mechanics. Proc. 4th Int. Conf. Soil Mechanics and Foundation Eng., London, 1, 116-119.

Tystovich, N.A. (1966) Permafrost Problems. In: Proc. Permafrost Int. Conf., 1963, Lafayette, Indiana. Nat. Acad. Sciences, Washington and N.R.C., p. 7.

Tystovich, N.A. (1975) The Mechanics of Frozen Ground. Scripta Book Co., New York. McGraw-Hill Book Co. 426 pp.

U.S. Army, Corps of Engineers (1950) Evaluation of Soils and Permafrost Conditions in the Territory of Alaska by Means of Aerial Photographs. Vols. I and II. Arctic Construction and Frost Effects Lab. Tech. Rept. 34.

U.S. Army/U.S. Air Force (1966) Arctic and Subarctic Construction; Calculation Methods for Determination of Depths of Freeze-Thaw in Soils. Tech. Manual TM5-852-6/AFM88-19. Chapter 6.

U.S.S.R. (1972) Guide for the Application of Drilling and Blasting Method of Loosening Frozen and Perennially Frozen Ground and Moraines. N.R.C., Ottawa. Tech. Translation 1877 (1976). 27 pp.

U.S.S.R. (1977) Atlas SSSR V Desyatoy Pyatiletke. Glavnoye Upravlaniye Geodezii i Kartografii pri Sovete Ministrov SSSR, Moscow. 60 pp.

U.S.S.R. (1980) Geografichaskiy Atlas dlya uchiteley sreney shkoly. 4th edition, Moscow.

Vallee, F.G. (1967) Kabloona and Eskimo. Res. Centre for Anthropology, Saint Paul Univ., Ottawa. 323 pp.

Van Cleve, K. and L.A. Viereck (1983) A Comparison of Successional Sequences Following Fire on Permafrost-dominated and Permafrost-free Sites in Interior Alaska. Permafrost. 4th Int. Conf. Proc. National Academy Press, Washington, pp. 1286-1291.

Van Everdingen, R.O. (1976) Geocryological Terminology. Can. J. Earth Sciences, 13, 862-867.

Van Everdingen, R.O. (1982) Management of Groundwater Discharge for the Solution of Icing Problems in the Yukon. Proc. 4th Can. Permafrost Conf., Calgary. N.R.C., Ottawa, pp. 212-226.

Van Everdingen, R.O. (1985) Unfrozen Permafrost and Other Taliks. Proc. Workshop on Permafrost Geophysics, Golden, Colorado. J. Brown, M.C. Metz and P. Hoekstra (eds.). U.S. Army, C.R.R.E.L., Special Rept. 85-5, pp. 101-105.

Van Everdingen, R.O. and H.D. Allen (1983) Ground Movements and Dendrochronology in a Small Icing Area on the Alaska Highway, Yukon, Canada. Permafrost: Proc. 4th Int. Conf., Fairbanks, Alaska. National Academy Press, pp. 1292-1297.

Viereck, L.A. (1965) Relationship of White Spruce to Lenses of Perennially Frozen Ground, Mount McKinley National Park, Alaska. Arctic, 18, 262-267.

Viereck, L.A. (1970) Soil Temperatures in River Bottom Stands in Interior Alaska. In: Ecology of the Subarctic Regions, UNESCO, Paris, p. 223.

Viereck, L.A. (1973) Ecological Effects of River Flooding and Forest Fires on Permafrost in the Taiga of Alaska. In: Permafrost. North American Contribution to the 2nd Int. Conf. on Permafrost, Yakutsk. Nat. Acad. Sciences, Washington, pp. 60-67.

Viereck, L.A. (1975) Forest Ecology of the Alaskan Taiga. Proc. Circumpolar Conf. on Northern Ecology, pp. I, 1-22.

Viereck, L.A. (1982) Effects of Fire and Firelines on Active Layer Thickness and Soil Temperatures in Interior Alaska. Proc. 4th Can. Permafrost Conf., N.R.C., Ottawa, pp. 123-125.

Vigdorchik, M. (1977) A Geographic Based Information Management System for Permafrost in the Beaufort and Chukchi Seas. Quarterly Rept. (April-June, 1977). U.S. Dept. of Commerce, National Oceanic and Atmospheric Administration (NOAA), Environmental Assessment Program (OSCEAP) Res. Unit 516, Contract 3-7-022-35127. 84 pp.

Vigdorchick, M.E. (1980) Submarine Permafrost on the Alaskan Continental Shelf. Westview Press, Boulder, Colorado.

Visser, S.A. (1973) Some Biological Effects of Humic Acids in the Rat. Acta Biol. Med. Germ., 31, 569-581.

Vorndrang, G. (1972) Kryopedologische untersuchungen mit Hilfe von Bodentemperaturmessungen (an einen zonalen Strukturbodenvorkommen in der Silvrettagruppe). Münchener Geog. Abh.,6. 70 pp.

Voroshilov, G.D. (1978) Effect of Coagulators on the Magnitude of Frost Heave of Far Eastern Supesses and Suglinoks. Proc. 2nd Int. Conf. Permafrost, Yakutsk, USSR. USSR Contribution. Nat. Acad. Sciences, Washington, pp. 261-264.

Vtyurina, E.A. (1974) The Cryogenic Structures in the Active Layers. (in Russian). Nauka Publishing House, Moscow, U.S.S.R. 126 pp.

Vyalov, S.S. (1959) Rheological Properties and Bearing Capacity of Frozen Soils. U.S. Army, Cold Regions Res. & Eng. Lab., Hanover, New Hampshire. Translation 74 (1965), 237 pp.

Vyalov, S.S. (1978) Long-term Settlement of Foundations on Permafrost. Proc. 3rd Int. Conf. Permafrost, Edmonton. N.R.C., Ottawa, 2, 297-312.

Vyalov, S.S. (1984) Placing of Deep Pile Foundations in Permafrost in the USSR. Permafrost. 4th Int. Conf. Final. Proc. National Academy Press, Washington, pp. 16-17.

Wang, W., J.W. Rooney and B.E. Davidson (1977) Simpson Hill Cut, Cooper River Basin, Alaska: a Case History of Slope Stability in Frozen Cohesive Soil. Proc. 30th Can. Geotechnical Conf., Saskatoon. Preprint. vol., pp. VIII-48 to VIII-74.

Washburn, A.L. (1947) Reconnaissance Geology of Portions of Victoria Island and Adjacent Regions, Arctic Canada. Geol. Soc. Amer. Memoir 22. 142 pp.

Washburn, A.L. (1956) Classification of Patterned Ground and Review of Suggested Origins. Geol. Soc. Amer. Bull., 67, 823-865.

Washburn, A.L. (1969) Weathering, Frost Action, and Patterned Ground in the Mesters Vig District, Northeast Greenland. Meddelelser om Grønland, 176. 303 pp.

References

Washburn, A.L. (1973) Periglacial Processes and Environments. E. Arnold, London. 320 pp.
Washburn, A.L. (1979) Geocryology - A Survey of Periglacial Processes and Environments. Edward Arnold, London. 406 pp.
Washburn, A.L. (1985) Periglacial Problems. In" M. Church and O. Slaymaker (eds.), Field and Theory. Lectures in Geocryology. Univ. of British Columbia Press, pp. 166-202.
Weaver, J.S. and J.M. Stewart (1982) In situ Hydrates Under the Beaufort Sea Shelf. The Roger J.E. Brown Memorial Volume, Proc. 4th Can. Permafrost Conf., Calgary, Alberta, March 1981, H.M. French, Ed., Associate Committee on Geotechnical Research, National Research Council of Canada, Ottawa, Ontario, pp. 312-319.
Webber, P.J. (1984) Terrain Sensitivity and Recovery in Arctic Regions. Permafrost. 4th Int. Conf., Fairbanks. Final Proc. National Academy Press, Washington, pp. 135-136.
Wechsler, A.E. and P.E. Glaser (1966) Surface Characteristics, Effects on Thermal Regime: Phase I. U.S. Army, Cold Regions Res. and Eng. Lab., Tech. Rept. 182. 91 pp.
Wein, R.W. (1976) Frequency and Characteristics of Arctic Tundra Fires. Arctic, 29, 213-222.
Wein, N. (1984) Agriculture in the Pioneering Regions of Siberia and the Far East: Present Status, Problems and Prospects. Soviet Geog., 25, 592-620.
Werenskiold, W. (1922) Frozen Earth in Spitsbergen. Geofysiske Publikationer, 2, 1-10.
Werenskiold, W. (1953) The Extent of Frozen Ground Under the Sea Bottom and Glacier Beds. J. Glaciology, 2, 197-200.
White, A.U. and C. Seviour (1974) Rural Water Supply and Sanitation in Less Developed Countries - a Selected Annotated Bibliography. International Development Res. Centre Rept. IDRC-028e. Ottawa.
White, S.E. (1976a) Is Frost Action Really Only Hydration Shattering? A Review. Arctic and Alpine Res., 8, 1-6.
White, S.E. (1976b) Rock Glaciers and Blockfields. Review and New Data. Quat. Res., 6, 77-97.
Williams, J.R. and R.O. van Everdingen (1973) Groundwater Investigations in Permafrost Regions of North America. In: Permafrost. North American Contribution to the 2nd Int. Conf. Permafrost, Yakutsk. Nat. Acad. Sciences, Washington, pp. 435-446.
Williams, P.J. (1967) Properties and Behaviour of Freezing Soils. Norwegian Geotechnical Institute Publication 72. 119 pp.
Williams, P.J. (1979) Pipelines and Permafrost. Longman, London. 98 pp.
Williams, P.J. (1984) Moisture Migration in Frozen Soils. Permafrost. 4th Int. Conf., Fairbanks. Final Proc. National Academy Press, Washington, pp. 64-66.
Williams, R.H. (1959) Ventilated Building Foundations in Greenland. Amer. Soc. Civil Eng., J. Construction Div., 85(602), 23-36.
Wittwer, S. (1984) The Rising Level of Atmospheric Carbon Dioxide: an Agricultural Perspective. In: The Potential Effects of Carbon Dioxide-induced Climatic Changes in Alaska. J.H. McBeath (ed.). School of Agriculture and Land Resources Management, Univ. Alaska, Fairbanks. Misc. Publ. 83-1, pp. 163-169.
Woo, M-K. (1976) Hydrology of a Small Canadian High Arctic Basin During the Snowmelt Period. Catena, 3, 155-168.
Woo, M-K. and R. Heron (1981) Occurrence of Ice Layers at the Base of High Arctic Snowpacks. Arctic & Alpine Res., 13, 225-230.
Woo, M-K. and P. Steer (1982) Occurrence of Surface Flow on Arctic Slopes, Southwestern Cornwallis Island. Can. J. Earth Sciences, 19, 2368-2377.
Woods, K.B. and R.F. Legget (1960) Transportation and Economic Potential in the Arctic. Traffic Quarterly, 14, 435-458.

References

Woods, K.B., R.W.J. Pryor and W.J. Eden (1959) Soil Engineering Problems on the Quebec North Shore and Labrador Railways. Amer. Railway Eng. Assoc. Bull., 60(549), 669-688.

Wright, R.K. (1981) The Water Balance of a Lichen Tundra Underlain by Permafrost. McGill Univ., Montreal, P.Q. Subarctic Research Paper 33. Climatological Res. Series 11, 110 pp.

Wu, T.H. (1984) Soil Movements on Permafrost Slopes Near Fairbanks, Alaska. Can. Geotechnical J., 21, 699-709.

Ye Bayou and Yang Hairong (1983) Determination of Artificial Upper Limit of Culvert Foundation in Permafrost Areas of the Qinghai-Xizang Plateau. Permafrost. 4th Int. Conf. Proc., Fairbanks. National Academy Press, Washington, pp. 1433-1439.

Yong, R.N. and J.C. Osler (1971) Heave and Heaving Pressures in Frozen Soils. Geotechnical J., 8, 272-282.

Younkin, W.E. and L.R. Hettinger (1978) Assessment of the Effects of Snow Road Construction and Use on Vegetation, Surface Elevations and Active Layers Near Inuvik, N.W.T. Proc. 3rd Int. Conf. Permafrost, Edmonton. N.R.C., Ottawa, 1, pp. 481-486.

Zarling, J.P., B. Connor and D.J. Goering (1983) Air Duct Systems for Roadway Stabilization Over Permafrost Areas. In: Permafrost: 4th Int. Conf. Proc. National Academy Press, Washington. 1524 pp. pp. 1463-1468.

Zhang Mingtao, et al. (1982) The Roof of the World. Harry N. Abrams, Inc., New York. 227 pp.

Zhao Yunlong and Wang Jianfu (1983) Calculation of Thawed Depth Beneath Heated Buildings in Permafrost Regions. In: Permafrost. 4th Int. Conf. Proc., Fairbanks. National Academy Press, Washington. 1524 pp. pp. 1490-1495.

Zhigarev, L.A. (1967) The Causes and Mechanism of Solifluction Development. (in Russian). Nauka. Moscow, U.S.S.R. 158 pp.

Zirjacks, W.L. and C.T. Hwang (1983) Underground Utilidors at Barrow, Alaska: a Two-Year History. In: Permafrost: 4th Int. Conf. Proc. National Academy Press, Washington. 1524 pp. pp. 1513-1517.

Zoltai, S.C. (1971) Southern Limit of Permafrost Features in Peat Landforms, Manitoba and Saskatchewan. Geol. Assoc. Can., Special Paper 9, pp. 305-310.

INDEX

Index

ABBREVIATIONS OF UNITS USED

°C	Degree Celsius
cm	Centimetre
d	Day
°F	Degree Fahrenheit
g	Gram
J	Joule
kg	Kilogram (10^3 grams)
km	Kilometre (10^3 metres)
kPa	Kilo Pascal (10^3 Pascals)
L	Litre
m	Metre
mg	Milligram (10^{-3} grams)
mm	Millimetre (10^{-3} metre)
MPa	Mega Pascal (10^6 Pascals)
mV	Millivolt (10^{-3} V)
N	Newton
°N	Degree of latitude north of the equator
°S	Degree of latitude south of the equator
V	Volt
yr	Year
µg	10^{-6} grams
µV	Microvolt (10^{-6} V)

THE PERMAFROST ENVIRONMENT

THE CROOM HELM NATURAL ENVIRONMENT –
Problems and Management Series

Edited by Chris Park, Department of Geography, University of Lancaster

THE ROOTS OF MODERN ENVIRONMENTALISM
David Pepper

ENVIRONMENTAL POLICIES: AN INTERNATIONAL REVIEW
Chris C. Park

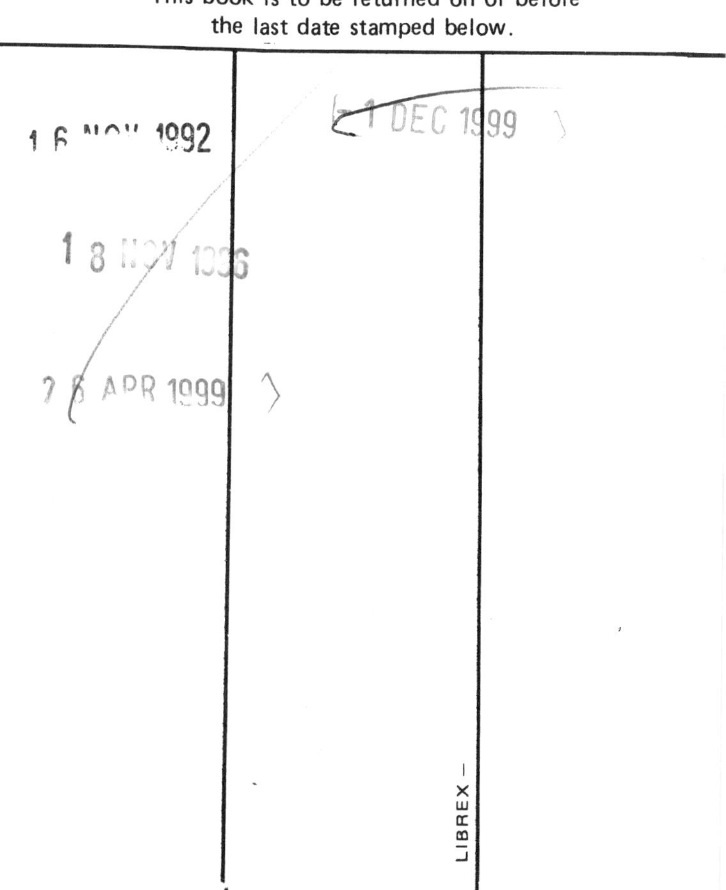
1 6 NOV 1992
18 NOV 1996
26 APR 1999
1 DEC 1999
LIBREX –